Management for the Construction Industry

CIOB Education Framework Textbook

Management for the Construction Industry

Stephen Lavender

Addison Wesley Longman Limited
Edinburgh Gate, Harlow
Essex CM20 2JE, England
and Associated Companies throughout the world

First published 1996

British Library Cataloguing in Publication Data
A catalogue entry for this title is available from the British Library

ISBN 0-582-26235-6

Transferred to digital print on demand 2001

Printed and bound by Antony Rowe Ltd, Eastbourne

Contents

Preface

The initial purpose of this book is to meet the requirements of the subject 'Management' in Level 2 Core Studies of the Education Framework of the Chartered Institute of Building. However, the structure and approach makes it suitable for similar studies in degree and diploma courses in the built environment disciplines. The book is pitched at an intermediate level. It builds on a likely prior knowledge of basic disciplines such as technology, economics and law, while forming the basis for more advanced studies in specialist aspects of management.

The main context of the book is the construction industry. However, emphasis will also be given to transferable skills in the study of management, particularly in the first three parts of the book — Foundations of Management; Functions of Management; Human Aspects of Management. Although these three parts will use construction examples where appropriate, the fourth part will apply management more specifically to the construction industry.

The book is intended to be comprehensive in its coverage of management, but its relationship to other texts is also of great importance. At the end of each chapter there is a guide to further reading. In addition, there is a bibliography at the end of the book.

Stephen Lavender
1996

Acknowledgements

I would like to thank my employers, the School of the Built Environment at the University of Glamorgan, for the support which enabled me to undertake the writing of this book.

I owe a particular debt of gratitude to my partner Kathy Tremain for her encouragement, and also for her very practical assistance in patiently reading the material and making numerous suggestions for improvements in content and style.

Part One

Foundations of management

1 An overview

Introduction

'Management' has an extremely broad practical application. It can include anything from handling a discontented bricklayer on site, to the planning of the multi-million pound capital budget of a major client. It is, of course, unrealistic to expect that the same person will carry out both functions, as they require very different skills. This gives some idea of the required breadth of a book designed to contribute to the education of managers in the construction industry.

The purpose of this book is to make a contribution by examining the principal concepts which inform management in general and are applied to the construction industry in particular. It is pitched at an intermediate level, that is, it assumes a basic technological, economic and legal knowledge. It is designed to form a basis for additional education and training appropriate to whatever aspects of management the reader aspires.

Management in society

Management has become one of the key words in society, with its use suggesting importance. Hardly a new concept or service is introduced these days without 'management' appearing somewhere in its label — total quality management and the management of change being but two examples.

Agreeing on a definition of the meaning of management is probably impossible since each individual will adapt the word to fit his or her own usage. Thus 'management' may be used to describe a complicated technological process, or it may be used to add weight to a much more straightforward process.

The implication is that the study of management cannot be objective, and although there are general principles which can be discussed, it is unlikely that a universal scheme can be devised which sets out the best way to manage.

The multi-disciplinary nature of management

Management is not a well-defined or single discipline in the sense that many academic subjects are. Instead it is best thought of as an area of interest which brings together aspects of many disciplines. Thus the study of management is informed by economics, sociology, psychology, engineering, law, statistics, and so on. Writers from all of these disciplines have made a contribution to the understanding of management.

The study of the subject of management is only part of the requirements needed to equip the reader to be a manager. Additional training programmes will be necessary, preferably undertaken throughout the working life, which develop the practical skills necessary to carry out management tasks, including the examples mentioned above — handling a discontented bricklayer on site, or planning the multi-million pound capital budget of a major client.

The meaning of management as used in this book

It will be apparent that since any organisation needs to be managed whatever its objectives, so management as a subject aspires to a degree of universality. This book is primarily concerned with management in the construction industry where most organisations are of a commercial nature, driven by economic objectives such as the need to make a profit. Commercial organisations of this kind are normally referred to as 'firms' in strict economic terminology, whereas management literature normally uses the term 'organisation' to emphasise the universal application of management principles. For the purpose of this book, the terms can be regarded as interchangeable, and although the main emphasis will be on commercial organisations, this will not preclude consideration of different organisations, or other objectives.

Outline of contents

As aforementioned, this book is a study of management as appropriate to the construction industry. It is designed as an intermediate treatment, building on relevant disciplines, and forming a basis for more advanced studies in specific areas. However, although the construction industry forms the principal context, to enhance the general applicability and transferable skills element of many of the principles, contexts of other industries and of the economy generally will be included.

Because of the wide-ranging content of the book, its relationship to other publications is extremely important. As well as references within the body of the chapters, there will also be guidance on further reading at the end of each chapter.

The main part of this chapter will provide an overview of the content of the book, drawing attention to the main principles to be examined. Following this, there will be a general guide to further reading in the study of management. The book is organised in four parts:

- Foundations of management
- Functions of management
- Human aspects of management
- Management applied to the construction industry.

Foundations of management

The seven chapters which form this part will identify key topics, particularly as they relate to commercial organisations. Many of the principles introduced in these chapters will form the theoretical basis of subsequent work.

Chapter 2 explores management perspectives, and recognises that it is difficult to be objective. There is no single view of management, or indeed of any of the social sciences. Historically, a variety of perspectives or schools of thought have developed. In management literature there are several, including 'Scientific Management', 'Human Relations', 'Bureaucratic', and 'Systems'. Industrial relations literature refers to 'Unitary', 'Pluralist' and 'Radical' perspectives, while economic commentators refer to 'Monetarist', 'Keynesian' and 'Radical' schools of thought. The purpose of this chapter is to explain the various perspectives — to show their similarities and differences. According to the perspective held by individual managers, different solutions may be implemented for a given management problem.

Chapter 3 examines the objectives of organisations. All organisations have the overriding need to survive. In the context of the commercial organisation, this can be translated into measurable objectives such as short-term liquidity to maintain cash flow, and long-term profitability for re-investment and growth. This will be explained in the context of construction organisations through an examination of revenue and costs, and by reference to the payments chain in construction projects. Beyond these basic requirements, individual organisations will vary in their objectives according to the perspectives of those in the organisation, and whether it is a private or public limited company. Some of the different types of construction organisation will be considered within the context of the changing structure of the industry.

Chapter 4 considers the external influences on organisations. Two major sets of external influences are market forces and public policy. Organisations have multiple market relationships, for example, as buyer of resources such as labour, materials, plant, subcontractors; and as seller of products such as houses, and the services of consultants. Capital must also be acquired through financial markets. Therefore the nature of competition and market strength must be considered, along with the plethora of commercial and contractual relationships found in the construction industry. An organisation's market position can greatly influence the achievement of objectives such as liquidity and profitability. Public policy on economic and legal matters can have a major impact on the management of organisations. This may occur in a general way through interest rates, or changes to employment or company law. Or it may have a more specific effect, say through health and safety legislation pertaining to construction.

Chapter 5 concerns internal influences within organisations. There has always been considerable debate on the extent to which all within an organisation share a common set of objectives. The debate often concerns matters of ownership and control, and is a reflection of the perspectives adopted, as discussed in Chapter 2. Among the matters to be considered are: whether there is a separation between ownership and control, and whether it matters; whether the task of management is to satisfy the needs of shareholders only, or some wider range of 'stakeholders'; whether there is inherent conflict between managers and other employees. This will enable an assessment of management in the organisation, and will be an important prerequisite for considering organisational structures in Chapter 7.

Chapter 6 concerns financial structures of organisations. An important requirement is to obtain sources of finance appropriate for particular uses. Short-term finance is required to provide working capital to maintain liquidity, while long-term finance is needed to invest in assets which form the basis for profitability. Organisational objectives and performance are often expressed in financial terms, for example through the balance sheet; profit and loss account; source and application of funds statements.

Chapter 7 examines organisational structures. Early studies of organisations were based on ideas such as bureaucracy — not a very appealing term, but a model to which many organisations still conform. The way in which an organisation's activities are structured vary according to a wide range of factors such as the type of industry, range of work, geographical dispersion, and how managers perceive organisational objectives will best be achieved. Consideration will be given to centralised and de-centralised structures; those based on functional department, and on multi-divisional structures. This chapter will also discuss matters such as hierarchies, spans of control, and communication within the organisation.

Functions of management

The five chapters which form this part will consider the various functions to be undertaken in managing the organisation, including marketing, production, finance and personnel.

Chapter 8 examines processes of management, and begins to study in more detail the practical aspects of the management role. This chapter is intended as a basis for Chapters 9–12. Management functions can be considered from the perspective of fulfilling organisational objectives, for example by increasing revenue and/or reducing costs, to improve profits. However, the traditional starting point for this topic is the work of Fayol, and his well-known list of management processes. This contrasts with the approach taken in Chapter 13 which instead considers what managers actually do, and how they spend their time. In addition, it is important to distinguish between general and project management, a classification particularly relevant to the construction industry.

Chapter 9 considers marketing management, one of the main purposes of which is to increase revenue and hence profit. It is usually argued that the contemporary firm is market orientated rather than production or operations orientated. This chapter includes a study of the main elements of marketing management, the organisation of the marketing function, plus what is known as the marketing mix — which products should be produced and for whom; what price should be charged; how the products should be promoted and distributed.

Chapter 10 concerns production management. Production creates the wealth of society by adding value to resources. For the firm, a major element of production management is the control of costs. The key to production management is the study of productivity, which is essentially a study of the efficiency of production. Consideration needs to be given to technological and social factors which influence production. A useful way of understanding production management is to note how it has developed over the years through the influence of the various schools of thought. This chapter will also examine some contemporary issues such as the influence of Japanese methods. In particular, ideas such as 'just-in-time' and 'total quality management' will be considered.

Chapter 11 concerns financial management. It was shown, in Chapter 6, how financial statements can be used to assess an organisation's performance. This chapter will further study the interpretation of these accounts, through financial ratios. Consideration will be given to the ROCE (return on capital employed) pyramid, which enables profitability to be analysed into profit margin and rate of asset turnover. Thus the financial structure and strength of organisations in different industries may be assessed. The apparent vulnerability of building contractors to cash flow crises is of particular interest. This chapter also considers financial planning and control, including the setting of different types of budget. An essential element of financial management is an analysis of costs, particularly the distinction between fixed and variable costs. This forms the basis for various management accounting techniques such as marginal costing, break-even analysis, standard costing, and long-term financial decisions such as investment appraisal.

Chapter 12 concerns personnel management. This is the final management function to be examined, and links with Part Three of the book. The importance of the personnel function and how it is organised will first be considered, followed by various aspects of personnel management. These include recruitment and selection; training; job evaluation and performance appraisal. The personnel function interacts with line management. This is problematical in the construction industry due to the remote location of many sites, and the preponderance of subcontracting and self-employment.

Human aspects of management

The six chapters which form this part will start with an examination of the skills required by, and behaviour of, managers. Consideration will then be given to individual aspects of human resource management such as motivation, followed by matters of group working and collective relationships at work. Finally, the legal framework affecting work will be considered.

Chapter 13 concerns management behaviour. It was shown in Chapter 2 that there are many schools of thought in the study of management. Therefore different managers have different perspectives, attitudes and assumptions which affect their behaviour. In particular, they will make different assumptions about the people they manage, and this will result in different management styles. Following from Chapter 8, there will be an examination of what managers actually do and how they allocate their time. This chapter will also consider the skills required by managers, including conceptual; human relations; administrative; technical.

Chapter 14 concerns leadership, which is of course a core management skill. For many, the word 'leadership' tends to suggest someone who is extrovert and 'up front', having been endowed with the gift of leadership rather than having acquired it. This perception needs to be questioned, because many commercial leaders are clearly not of this type. Different leadership styles may be appropriate, depending on the job. For example, a group of highly motivated professionals may simply require some reflective co-ordination. There is a good deal of literature on theories of leadership, for example, trait, style, contingency theories, as well as the 'best fit' approach. In addition, leadership will be examined in relationship to the culture of the organisation.

Chapter 15 concerns individual motivation, which is one of the most widely discussed topics in management literature. There is great diversity of view between the various schools of management thought, mainly stemming from the different assumptions made about people and their needs. For example, scientific management is often associated with purely economic methods of motivation. There has also been a major contribution to motivation theory from the discipline of psychology, for example through the work of Maslow and others. Indeed the 'psychological contract' between employer and employee is an important concept in assessing whether the expectations of both parties are being met.

Chapter 16 is concerned with work groups. It is very useful to examine individual motivation, but of course work is often a collective or group activity. Indeed, some motivation theory recognises the benefits of social contacts in the workplace, whether they are through formal or informal groups. An important early study in this field was the Hawthorne experiments, which have influenced much subsequent work. Various aspects of group behaviour will be considered, including their development and effectiveness. Work groups may be permanent, or temporary. Of particular relevance in the construction context is the study of the operation of project teams, and how the various disciplines may be blended to produce a coherent whole.

Chapter 17 considers industrial relations. The starting point is to reiterate the industrial relations perspectives, first introduced in Chapter 2. The participants in the industrial relations system include trade unions and employers/employers associations, who undertake collective bargaining at a number of levels. This is of course heavily influenced by the labour market, the workings of which will be studied. Also to be discussed will be the high profile issue of industrial conflict. It is also important to consider how the conduct of industrial relations has changed in recent years, in response to changes in economic conditions and government policy. This will lead into the legal aspects discussed in the next chapter.

Chapter 18 considers the legal regulation of work. Employment law plays an important part in regulating employment relationships, both individual and collective, and this chapter will summarise the main elements. Most employment law is civil rather than criminal, and is a mixture of common law and statute. A convenient method for studying the law is to divide it into individual and collective law. Individual law is based on the contract of employment, and the main elements to be discussed are the contents of the contract, and the grounds for termination. A particular legal problem for construction arises from the widespread practice of self-employment. In collective employment law, the main issues concern collective bargaining rights, and rights in respect of industrial action. In all cases, both common law and statute law need to be considered. More recently, European law has also made an impact. As in Chapter 17, the changes which have occurred in recent years will be assessed.

Management applied to the construction industry

Although the first three parts of this text will relate to construction where appropriate, considerable emphasis will be placed on transferable skills. By contrast, the six chapters which form this part will use the framework of Parts One, Two and Three to discuss particular management problems related to construction organisations and projects. In this text, they will be dealt with in principle and outline, to form a basis for further consideration at more advanced levels of study.

Chapter 19 considers market strategies for construction firms. This is a key issue for contemporary market-orientated firms and involves researching the market, and putting in place appropriate organisational and financial structures. There have been great changes in recent years in terms of market structure and market strategy in construction. On the one hand, there has been the emergence of a relatively small number of large, diversified national contractors, and on the other hand, of large numbers of smaller subcontractors and specialists. Consideration will also be given to the demand conditions which firms face, in terms of quantity, quality and methods of delivering demand. As a result of studying the structure of the industry, the shape of demand, and the availability of finance, the market opportunities available to the construction firm can then be considered.

Chapter 20 continues the study of construction firms by examining certain important policy matters. The market-orientated firm must develop organisational structures which are responsive to changes in market conditions, and this will be considered first. Following this, three areas of policy making will be examined. Plant management is a traditional concern of firms, and questions such as whether plant should be bought or hired, and what should be the function of the plant department, will be considered. The question of quality management is of more recent concern. It was seen in Chapter 10 how this concept has been introduced into manufacturing. This chapter will discuss the applicability of quality management to construction. Health and safety is another policy area of continuing concern. This issue has to be carefully considered by firms for many reasons, including recent changes in the legal framework.

Chapter 21 changes the emphasis from the management of firms to the management of construction projects. The main elements of project management will be discussed, with a particular emphasis on satisfying client needs. The focus will then shift to project planning and control — why planning is needed, and the process of planning. There are various planning tools and techniques available to the manager, the two most common of which are bar charts and networks. Project planning is usually associated with time management, but consideration will also be given to the relationship between time and cost. Finally, the availability and usefulness of project management data will be discussed.

Chapter 22 considers financial control of projects. Effective financial control lies at the heart of fulfilling the objectives of the organisations participating in the project. It is a constant theme of this book that commercial organisations require profitability and liquidity if they are to survive and prosper. Therefore this chapter will explain the problems for both clients and contractors in meeting these objectives, and how these problems may be overcome. In the construction process much depends on the outcome of the series of market transactions between the various parties. Each market relationship is regulated by a commercial contract, and the whole series of relationships link together into the payments chain. Related specific topics to be discussed in this chapter include the reasons why the agreed contract sum may be altered, and the problems of liquidity for both client and contractor.

Chapter 23 concerns production management in construction. This too involves cost control, but whereas the previous chapter dealt with costs incurred through market transactions, this chapter examines costs incurred in the production process through the use of resources. Following a historical survey of the development of production methods in construction, the focus will be on productivity in construction, and the various factors which influence productivity performance. Finally some techniques for improving productivity, such as work study, will be briefly considered.

Chapter 24, the final chapter, considers future prospects for the management of construction. The management of change is a constant theme of this text, and the general underlying principles will be reviewed. Three particular areas of change will then be examined. Firstly, changes to the construction market which can be expected in the last few years of the century. Secondly, the way the industry conducts its affairs. The industry has often been criticised for being too conflictual and for performing poorly for its clients. There has been considerable debate about how the situation should change, with the report *Constructing the Team*, published in 1994, acting as a focus. Thirdly, there is public concern about the environment. Just about everything the construction industry does affects the environment, from how land is used through to the energy consumed and pollution generated by the production of materials.

Guide to further reading

The importance of further reading cannot be over-emphasised. In the course of most chapters, reference will be made to other sources of information. At the end of each chapter there will be a section giving advice on further reading. Furthermore, it is appropriate in this introductory chapter to list some of the publications which will be useful at several points in this text. In all these cases, the author and the year of publication will be given, with full details appearing in the Bibliography.

Having identified management as being substantially concerned with the achievement of economic organisational objectives, it is important for the reader to have a sound grasp of economic principles. For this, see Lavender (1990) and Neale and Haslam (1994).

There are numerous books of a general nature on management. Those of a UK origin include: Dixon (1991), a straightforward summary of the main principles; Cole (1993), more detailed and comprehensive; and Handy (1993), a well-established and individual management text by an eminent writer. There are many very detailed and comprehensive management books of US origin, including: Megginson *et al.* (1992) and Koontz and Weihrich (1990). There are also several books which relate management to construction, for example: Calvert *et al.* (1995); Fryer (1990); Forster (1989); Harvey and Ashworth (1993); Hillebrandt and Cannon (1989); and Walker (1989). Others will be referred to in the various chapters and listed in the Bibliography.

In addition to books, reference can also be usefully made to periodicals such as *Building*, and *Chartered Builder*, and to quality newspapers such as *The Guardian*, *The Independent*, *The Daily Telegraph* and *Financial Times*.

Many other publications, books and articles, will be specifically referred to in conjunction with particular chapters, and listed in the Bibliography.

2 Management perspectives

Introduction

Since there is no universally accepted view of the correct way to manage, management cannot be regarded as an objective science. Different organisations, circumstances and management tasks require different approaches. Furthermore, individuals involved in the management process bring to their work a variety of intellectual positions, as well as cultural and value systems, all of which affect their behaviour.

The apparent difficulty in achieving objectivity is a problem shared by many of the disciplines from which the study of management is drawn, particularly the social sciences. This is because the study of people in society, and in organisations, is influenced by a range of social, political and economic factors. Therefore a range of views exist on the meaning of management and the nature of the manager's job.

The purpose of this chapter is to explore some of these diverse views. In the development of management thought, a number of perspectives, or schools of thought, have developed. However, given that management is primarily concerned with achieving economic organisational objectives, it is appropriate to first consider different views of the economic system. As will be seen, major differences exist on attitudes to the market as a means of organising economic affairs, and these find echoes in the schools of management thought, especially in the industrial relations literature. Many of the themes introduced in this chapter will form a basis for much subsequent material.

Economic analysis and management thought

A key area of economic debate which has emerged since the late 1970s, and which still dominates, is the role that should be played by the market. For most of the post-1945 period up until the late 1970s it was widely accepted that the free market would not produce the best results, and that there was therefore a need for substantial public intervention. This view was broadly accepted by all governments, and as a result there was some consistency in economic policy throughout the post-war period.

This approach was reflected to some extent in management theory and practice, where reliance on the market for determining wages and conditions of work was amended by collective bargaining and moves towards industrial democracy. Thus the role of management was not simply to maximise profits for the owners of the firm, but to satisfy a wider range of interests, including those of the employees.

This line of thinking has largely been replaced in more recent years, in terms of both economic policy and management practice, by reliance on the rule of the free market. Thus the state has a much smaller role in running the economy, while management policy has shifted away from collective bargaining towards implementation of market-determined wages, maximising of shareholder benefit, and asserting the right of management to manage.

However, it would be wrong to assume that there is unanimous agreement that this is the best approach. There are **three** well-established schools of economic thought, which may be broadly summarised as:

1. *The market is a good mechanism,* which should be encouraged by removing barriers to its efficient operation.
2. *The market is a reasonable mechanism* in many respects, but does require a degree of public intervention and regulation to overcome its faults.
3. *The market is a poor mechanism,* which results in wasted resources and injustices, and ways should be found to replace it wherever possible.

Each of these views will be considered, together with the parallel strands of management thought, particularly as expressed in the industrial relations literature.

'The market is a good mechanism' view

This view has been the basis of government policy since 1979. The underlying economic philosophy is known as **monetarist**. It is believed that market forces are best able to achieve efficiency and consumer choice, and that obstacles to the free market hinder this process. Therefore the government encourages 'free and flexible' markets for labour, capital and goods.

The implications for management include reliance on the market for setting wages and conditions of employment; individual contracts of employment rather than collective bargaining; working within the disciplines of global markets. Under these arrangements the interests of shareholders, as expressed through free capital markets, are paramount. The interests of workers are identical to those of the firm, and will be enhanced by its success. Thus there is no room for industrial conflict, which is regarded as illegitimate.

In the industrial relations literature this viewpoint is referred to as the **unitary** frame of reference, and is undoubtedly adhered to by large numbers of managers, particularly at a senior level.

'The market is a reasonable mechanism' view

This view was widely held until 1979, and has a much stronger tradition in European countries other than Britain. The underlying philosophy, in Britain at any rate, is referred to as **Keynesian**, named after the economist John Maynard Keynes, whose ideas, developed in the 1920s and 1930s, gave intellectual force to this view. This is a 'mixed economy' approach which recognises deficiencies in the market such as monopoly power, inequity, and the inability of the free market economy to sustain high levels of output and employment. Therefore there is an important role for government in correcting, or even replacing, the market where necessary.

A major implication for management is that there are a range of interests to be considered as well as those of shareholders. Indeed some conflict of interest is quite likely, say, between management and workers. However, it is believed that conflict can be resolved by an appropriate balance of power and the use of correct procedures. Thus there is an important role for collective bargaining and even industrial democracy within the organisation.

This is referred to as the **pluralist** frame of reference in industrial relations. Although this viewpoint tended to operate in many firms up until 1979, it is not known how many managers believed it was best, rather than merely going along with it because of prevailing economic conditions. Possibly only a minority of British managers would support this approach today.

'The market is a poor mechanism' view

This radical view holds that for the main sectors of the economy the market is a poor method of organisation. This approach is often referred to as **Marxist**, named after the nineteenth-century social scientist, Karl Marx, who analysed the workings of capitalism, particularly in Britain, and founded a school of thought which is still influential. It is argued that the benefits of the market system, for example as propounded by contemporary monetarists, are an illusion. This is because the free market is subject to periodic crises of under- or over-production leading to waste and inequity. Furthermore, it is not possible to tamper with the market in the Keynesian sense since this will only give short-term respite before the 'logic of the market' reasserts itself. In this latter respect there is a good deal of similarity between the Marxist and monetarist views.

In the industrial relations literature, this viewpoint is referred to as the **radical** frame of reference. If the Marxist view is an accurate assessment of the way the market works, then there is a good deal of inevitable deeply rooted conflict in organisations. Furthermore, this conflict is based on the notion of class struggle, since the interests of labour and capital are diametrically opposed. While few managers are likely to subscribe to this view openly, it is possible that the behaviour of some may indicate support for this approach. Certainly this is a view held by some workers and/or their leaders, and by those politicians who regard striking workers as 'the enemy within'. Perhaps this is another example of how the monetarist and Marxist analyses have a tendency to converge.

Management schools of thought

The previous section shows how the main strands of economic thought affect management. Since the study of management is principally concerned with economic objectives of organisations, it might be expected that the development of management thought has followed a similar path. This is not quite the case. Therefore the conventional classification of management schools of thought will be explored, prior to considering their relationship to economic thinking.

Management as an art has always existed, and aspects have been documented since the beginning of the industrial period. For example, in what is usually regarded as the first book on economics, *The Wealth of Nations,* Adam Smith memorably describes production of pins using the division of labour. However, it was not until the late nineteenth century that management started to become formalised as a specific study, at which time concepts such as the division of labour were very much to the fore.

It is generally accepted that the founding father of modern management thought was F.W. Taylor, who coined the phrase 'scientific management', the label given to the oldest school of thought. This is a rather production-orientated technologically based viewpoint with units of labour feeding into the production process in a somewhat passive manner. Later schools of thought gave more significance to the human element in production. In the post-1945 period, schools of thought have considered a wider variety of functions and contexts in which the organisation operates. There is no general agreement on the classification of the various schools of thought, and textbooks often take slightly different approaches. However, it is convenient to classify as follows:

- scientific management
- human relations
- post-1945 approaches.

Scientific management

The aim of scientific management was to identify universal principles on which production could best be organised. It was assumed that there was a 'best way' of doing things, and it was the task of management to determine what it was. Work could be analysed into its simplified component parts, whereupon it would be a relatively straightforward matter to hire, train and pay the necessary labour. This approach has similarities with the traditional economic theory of production. This assumes fixed capital in the short run, to which varying amounts of homogeneous labour can be applied, having been hired, at the market wage, through free and flexible labour markets.

In attempting to interpret the scientific management school, there are, however, some differences of approach. For example, it is often thought that by finding the best way of producing, productivity will rise and so workers should receive better wages. This is the basis of the argument that scientific management uses money as a motivator for workers. However, a more radical interpretation is that scientific management aims to mechanise production as much as possible so that the work becomes de-skilled. By this process each 'unit of labour' is much the same as another, so the bargaining power of each worker is reduced, allowing wages to be pushed down as low as the market will bear. The outcome of this approach will be influenced by economic conditions, especially the level of unemployment.

It is appropriate to identify some of the main figures in the scientific management school of thought, and outline their contributions, most of which date from the 1920s.

F.W. Taylor

He was the father of scientific management, who had a major interest in the efficiency of manual work, and how this might be improved so that productivity, profits and wages could be higher. He might be regarded as the original 'downsizer' because, as a result of his studies of work at Bethlehem Steel in the USA, the workforce was cut from 600 to 140. However, those who remained received 60 per cent higher wages as long as they met their work targets.

F. and L. Gilbreth

Frank Gilbreth was a bricklayer who took part in some of Taylor's studies and eventually continued this into more detailed work on time and motion study. He worked closely with his wife Lillian, who in addition began to recognise how environmental factors such as heating and lighting affected the performance of workers.

H. Gantt

Henry Gantt is best known for having developed some of the principles on which time management is based. The familiar bar chart used on every construction site is often called a Gantt chart, as he is credited with its invention.

H. Fayol

Unlike the above mentioned, who were American, Henri Fayol was a French industrialist. A significant difference was that whereas the others tended to concentrate on the production process and productivity, Fayol was concerned with the broader role of management. He is known for his attempt to develop universal principles through which all organisations could be managed. Fayol's functions and principles of management are still a central part of the study of management.

H. Ford

With only Taylor as a possible rival, the name of Henry Ford most epitomises scientific management. In fact twentieth-century large-scale assembly-line industrial production is sometimes referred to as Fordism, while more modern flexible production strategies are sometimes called post-Fordism.

Human relations

The human relations school developed mainly in the 1920s and 1930s. According to Megginson *et al.* (1992), the springboard for the human relations school of thought was Oliver Sheldon's statement of a philosophy of social responsibility in 1923. This identified a responsibility for management beyond the scientific management imperative of maximum efficiency. However, the ideas were not entirely new, since Robert Owen had put some of them into management practice in the early days of the Industrial Revolution. Furthermore, nineteenth-century writers such as Marx and Durkheim had warned of the potential de-humanising consequences of industrial production organised along the lines of scientific management, through their concepts of, respectively, *alienation* and *anomie*.

The best-known studies contributing to the human relations school were those carried out in the USA at the Hawthorne plant of Western Electric, under the direction of Elton Mayo. These studies were also about increasing productivity and efficiency, but found that motivation was greatly enhanced when workers felt that they were part of a group. Therefore, social interaction was seen as beneficial. One of the findings of the Hawthorne studies was that informal organisations develop within the formal. Management should not necessarily discourage this, since it can result in better performance.

Human relations goes beyond scientific management in the sense that people are not regarded as mere machines who only have to be paid a satisfactory rate to motivate them to work. However, it would appear that workers are still regarded as 'units of labour' in that they are still subject to direct management control and have little say in how the organisation and its work should be carried out. Developments in management thinking which gave more responsibility to workers occurred later, in the post-war period.

Post-1945 approaches

This period has been characterised by a diversity of schools of thought, some of which overlap. This somewhat confusing situation may have arisen because there is much greater interest in management studies across a wide cross-section of academics and practitioners. The greater diversity of approach has also arisen because, whereas the scientific management and human relations schools were somewhat one-dimensional, some of the post-1945 views are more ambitiously multi-dimensional. Even where the work of the traditional schools have continued, they have become more varied, such as in the field of motivation theory. Two of the more influential schools of thought to be briefly considered are:

- systems theory
- contingency theory

Systems theory

The systems approach is certainly multi-dimensional. A system has a large number of parts which interact with each other. Each part is a subsystem of the whole. Therefore a whole system, such as a commercial organisation, will operate according to how the individual elements such as the technology, people and management style interact.

Most organisations are *open systems* in that they interact with their environment. Thus the organisation as a system takes in resources from the outside environment, processes them and then sends them out. The external environment with which organisations often interact are resources markets from which labour, materials and capital are bought; and product markets to which the organisation's output is sold after the resources have had value added to them. Systems which do not interact with an external environment are referred to as *closed systems*.

In addition to the classifications of open and closed systems, it is also useful to think of the *degree of openness*. The contemporary firm is often described as market orientated. This means a higher degree of openness in that it looks outwards to its existing and potential customers in order to determine the exact demand from the market. Thus a system can then be designed which best meets that demand. This contrasts with a production-orientated system which has a lower degree of openness. In this case the organisation will design its system first before looking outwards.

Contingency theory

This approach is concerned with adapting to change, since it recognises that no universal approach to management is possible. The most appropriate form of management will differ between organisations, and will differ over time within the same organisation. The similarity with systems theory is that there is some emphasis on the environment in which the organisation exists. In the case of contingency theory, the environment is one of the major influences for change.

It would appear that contingency theory can embrace many of the other schools quite comfortably by simply recognising that each situation must be considered separately. Within this general umbrella there are particular lines of enquiry. For example, some studies have concentrated on the effect which the prevailing technology has on organisational structure and management style. Assembly-line production requires bureaucratic structures, whereas individual construction projects require temporary teams. Generally speaking, the faster changing is the environment, then the more flexible should be the organisation.

Management perspectives today

Can it be said that the schools of management thought have much relevance today? It is often believed that scientific management is of an earlier era, and is now dead and buried. In the view of some, this has never been the case. For example, there is a strain of radical literature starting from Braverman (1974), which argues that scientific management, or 'Taylorism', has always been the preferred option of industrial managers. The basis of the argument is that scientific management techniques, such as the division of labour and mechanisation of the production process, are best for higher productivity, this being the key to reduced costs and higher profits. It is further argued that the more human relations type approaches were only widely adopted by management when forced to do so by particular economic conditions. Therefore, in periods of low unemployment, workers are able to resist scientific management, and demand a more humane approach.

Of course this view is open to criticism, even from other radical sources. There may be agreement that management will always seek to maximise productivity and profits to the detriment of workers. However, this can be achieved through a variety of methods, including human relations type approaches. This variable approach has something in common with contingency theory.

If management approaches in recent years are considered, it could be argued that economic circumstances play a very influential role. In the 1980s and 1990s high unemployment has seen a shift away from the people-based approaches in many organisations towards more assertive management approaches. (Perhaps 'macho management' should be added to the role call of management schools of thought!) However, perhaps it is wrong to generalise. Rather, what has been seen is the establishment of a diversity of approaches where different firms use different approaches according to their own specific circumstances.

Summary

The main theme of this chapter is revealed by its title — there is no single perspective which explains the theory and practice of management. On the assumption that the main context of this book is the management of firms — that is, commercial organisations driven by economic objectives — it was appropriate that the economic context was first examined. It has been shown that a major issue is the extent to which the market determines the behaviour of organisations. The diversity of view finds echoes in the industrial relations frames of reference. However, links with the general management literature are less clear. While the scientific management and human relations schools sit quite comfortably with market theories, the schools of the post-war period are more diverse, reflecting the pluralism of the times. Nevertheless, economic conditions have been very influential, and the high unemployment of much of the period since the late 1970s has undoubtedly had an impact on management theory and practice.

This chapter has attempted to give a flavour of the great variety of management perspective. It would seem reasonable to conclude that each organisation will devise strategies which are appropriate in its own circumstances. These may change over time, and may indeed vary for different parts of the organisation. This variety of perspective is fundamental to the study of management and the material in this chapter will frequently be referred to in the remainder of this book.

Further reading

There is, of course, an enormous amount of published material on the subject matter of this chapter. For the economic schools of thought refer to Lavender (1990), chs 1 and 3. For industrial relations frames of reference, see Jackson (1991), ch. 1. For the schools of management thought, see Dixon (1991), ch. 1; Handy (1993), ch. 1; and Megginson *et al.* (1992), ch. 2. For a radical interpretation of scientific management, see Braverman (1974).

3 Objectives of organisations

Introduction

Most people, on being asked to name the objective of a commercial organisation such as a firm, would be likely to say that it is to make a profit. This is understandable because profit is a well-defined variable which is capable of measurement. However, this is somewhat simplistic because everything is being reduced to a single variable. Furthermore, although 'profit' can be and indeed needs to be considered objectively, it is often controversial. For example, when firms make large profits there may be public debate on whether it is justified. Sometimes the debate may become heated, possibly with political implications.

An alternative approach is to assume that the overarching objective of an organisation is to survive. After all, without survival those working in the organisation would find their livelihoods threatened. When considering the essential requirements for the organisation to survive, it becomes clear that most of the variables are difficult to measure. The exceptions, of course, are financial targets such as profit. But profit is not the only financial necessity, since a potentially profitable organisation could still fail to survive, due to a lack of liquidity. This is a very familiar problem in the construction industry, a consequence of the complicated contractual relationships in the construction process, which will be illustrated by reference to the payments chain.

It is all too easy to discuss the organisation as if it were an impersonal entity, but of course it contains people. Thought must therefore be given to whether different groups of people within the organisation may modify its objectives from any which are stated. In particular, it must be considered whether pressures exist which result in an organisation pursuing objectives other than profit maximisation.

This chapter will consider the principles on which organisational objectives are founded. The next two chapters will expand on some of these principles.

Explaining organisational objectives

It is common in management books to study objectives of organisations in the context of the function of planning, as for example in Dixon (1991), Cole (1993) and Megginson *et al.* (1992). Thus the setting of objectives is part of the planning process. This will be considered in more detail in Chapters 8 and 21, but it is sufficient to say at this stage that planning occurs at a number of levels in a number of time-scales.

In respect of time-scale, for example, Dixon (1991) argues that there are two types of objective — those which define the purpose of the organisation now, and those which set out its proposed purpose in the future. In some cases, an organisation may envisage unchanged objectives in the future, whereas others may envisage growth into new markets. This distinction seems particularly relevant when considering the position of the now privatised but previously nationalised utilities. While in public ownership, the main objective of such an organisation might be to provide a service to the public. However, as a private company the main objective may be to ensure that the interests of shareholders are served.

Many writers do not entirely agree with the assertion that the main objective of a firm is to pursue profits. For example, Handy (1993) claims that this is too simplistic because it is not known

- how strongly
- over what time-span
- with what degree of risk and
- within what constraints

profit is pursued. Thus Handy argues that it is more useful to regard profit as something which is necessary for achieving a range of objectives such as survival, growth and market share.

Another way of defining organisational objectives is to classify firms as

- production orientated or
- market orientated.

A production-orientated firm is one in which decision making starts from what it believes it can profitably produce. This may be based on the productive capacity of the firm's resources, such as the optimum output of its factory. Only then does the firm pay attention to potential purchasers of the products, and how the products might be marketed. By contrast, a market-orientated firm is one which studies its market first to decide what customers require. Production is then geared around these market requirements in the most cost-effective way.

It is widely believed that the successful contemporary firm must be market rather than production orientated, thus emphasising the needs of customers. This will be more fully explored in Chapter 9 on marketing management. However, it should be noted that writers such as Galbraith (1972) have argued that this underestimates the power of large firms, especially those operating transnationally, to decide what they can produce most effectively, and then manipulate the market to secure sales. The ability to achieve this may depend on economic conditions.

Organisational survival

As previously stated, a useful starting point may be to assume that the overarching objective of an organisation is to survive. Thus in Drucker (1972), and in other publications by the same author, the survival needs of the firm are explored. **Five** areas of survival objectives are described. A minimum standard must be reached in each:

1. There must be an *effective human organisation* within which people work as individuals, and also as a group, for a common result. Furthermore, this human organisation must be capable of perpetuating itself beyond the life span of any one management group. This is because many decisions are long term in nature and might take a considerable time to bear fruit. Therefore, the cult of the individual, charismatic entrepreneur may be useful to inspire a firm in the initial stages of its life, but can give rise to succession problems later on.
2. The firm exists in *society and the economy*, and must respond to them if it is to survive. This means that the firm must be aware of public opinion and government policy, and work within their demands. There must be a thorough understanding of how economic policy affects the management of the organisation. For example, a firm ought to assess the likelihood of the government altering interest rates, and the effect this would have on the organisation. Another important issue for the latter part of the 1990s might be the effect on the firm's activities of growing public concern for the environment.
3. The firm exists to *supply an economic good or service*. This is the most obvious reason for a firm to exist as far as the public is concerned. In some respects this is an extension of point 2, because one of the aspects of society and the economy which the firm must respond to is to satisfy market demand by producing goods and services which people demand.
4. The firm must *adapt to change*, or, better still, innovate. The kind of changes which affect organisations could be economic, technological or social. For a firm to prosper, it must be capable of innovation. Drucker believes that the business enterprise is the one type of organisation specifically designed to innovate.
5. *Profitability* is the absolute essential for survival because it allows for risk. Making an allowance for risk may not necessarily represent a cash flow out of the firm, but it nevertheless represents a significant cost. If risk is not allowed for then assets cannot be replaced, and the capital base of the organisation and its capacity to produce will be eroded.

Drucker's analysis has been influential, but can be criticised on the grounds that it takes a view of firms which does not allow for any conflict. Therefore, within the definitions of Chapter 2, a unitary perspective has been adopted. Furthermore, although it might be difficult to argue against the proposition that firms need to make a minimum profit in order to re-invest and survive, let alone grow, profits are not necessarily quoted in this way. A firm's profit, as identified by accounting procedures, show the total figure including any sum which is gained, for example, through the exercise of monopoly power. Thus, it is not easy to separate out the 'legitimate' profit necessary for re-investment from the 'illegitimate' profit derived from monopoly power.

Before leaving the question of survival, it is worth questioning the assumption that everyone concerned with the organisation will have the overriding desire that it should survive. Surprisingly enough, it is possible to argue that the owners of the firm — that is, the shareholders — may not feel that it is in their best interests for the organisation to survive, or at least not as an independent entity. A possible scenario is that a competitor may make a takeover bid for the firm by offering the shareholders a tempting price for their shares. This competitor's intention may be to rationalise or close down the firm, so if the shareholders accept the offer, then survival of the firm is threatened. To understand this argument more fully the effect of the capital markets on the firm will be examined in the next chapter.

Profit

This concept recurs throughout this chapter. Some argue that the job of management is to maximise profits for shareholders, while others maintain that profit is one of several survival functions which have to be fulfilled. The common ground is that some level of profit is necessary for the firm to guarantee survival in the long term. The next section will examine the meaning of profit in more detail.

The objective need for profit

For a firm to commence operating there must be an input of resources to initiate the 'cycle of production', as shown in Figure 3.1. This is an economic process comprising the three stages:

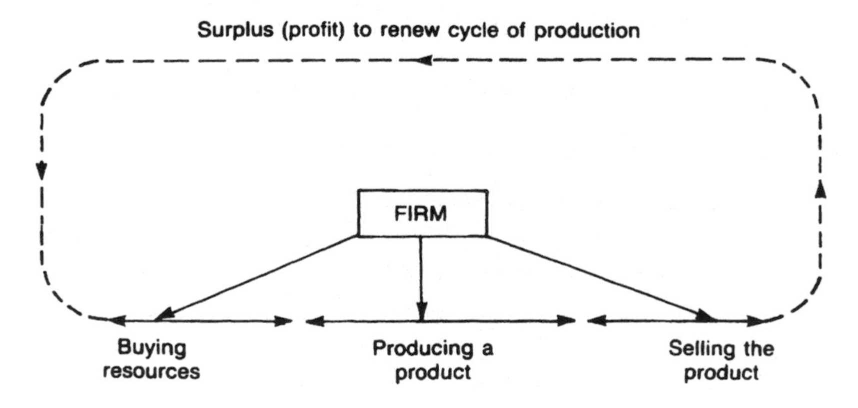

Fig. 3.1 The cycle of production

- buying resources
- producing a product
- selling the product.

In this process the firm buys resources out of capital which has been made available by the owners or lenders. The resources are broadly of two types:

- **variable**, meaning those wholly used up in the cycle of production, such as materials and labour time, and
- **fixed**, meaning those only partly used up, such as plant and machinery.

On completion of the production cycle, the process must be renewed. This is a prerequisite for the firm's survival. Two further inputs of capital are required in order to

- buy more variable resources, and
- allow for depreciation of the fixed capital so that it can be replaced when it is eventually used up.

These further inputs of capital are essential requirements if the firm is to remain intact, let alone grow. Therefore, the firm has to generate a surplus from its activities in order to guarantee its survival. This, of course, is a description of profit as the objective need of the firm. If a firm does not generate a surplus at first, then further capital might be injected by the owners or through borrowing. However, there must be a prospect of surplus or profit in the longer term so that a return on capital is achieved.

The components of profit

Having established an objective need for profit, it is important to investigate how it is achieved. Profit comprises two separate components which are in large measure independent of each other. Thus to make a profit, a firm must satisfy the following conditions:

- it must produce effectively by acquiring resources for the lowest price possible, and then using those resources as efficiently as possible; this controls **costs**
- it must sell effectively in the market by finding the right balance between price charged and quantities sold; this generates **revenue**.

Thus, the final profit is an arithmetically derived figure dependent on the amount of revenue generated from sales, and the level of costs incurred in production. Therefore,

PROFIT = REVENUE – COSTS

The way in which profit is derived can be shown in a pyramid form, as in Figure 3.2. This analysis of profit can be used to show how costs and revenue are themselves derived.

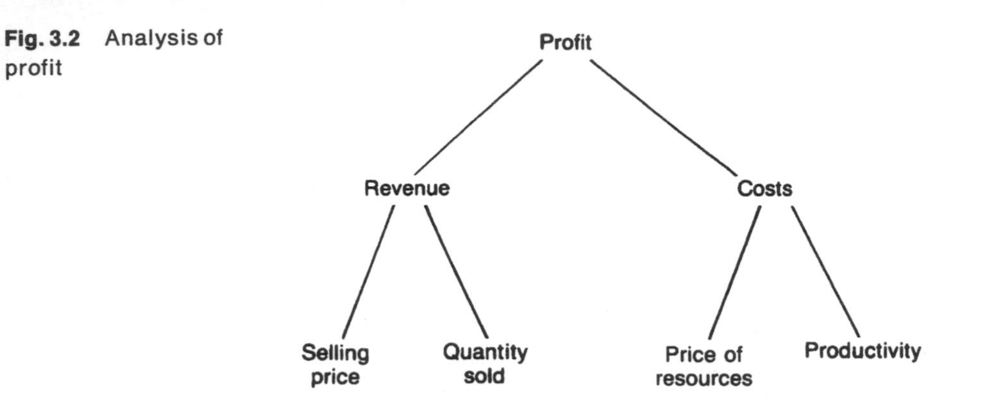

Fig. 3.2 Analysis of profit

Costs

Costs are incurred according to prices paid for resources and the efficiency with which those resources are then utilised. This measure of efficiency is usually referred to as productivity. Often there may be a trade-off between the price of resources and productivity, since the cheapest resources may not perform as well, and so overall costs may be higher. For example, cheaper materials may have a higher wastage factor or be more difficult to work with, thus leading to a productivity penalty and higher costs than would have been the case with more expensive, better quality materials. Similarly, lower wage rates may imply a labour force with lesser skills who therefore require more costly supervision to prevent poor or even abortive work.

Revenue

There may be a trade-off similar to above in that if the selling price is increased, quantity sold may decrease to the extent that total revenue is reduced. Thus there is normally an inverse relationship between price and quantity sold, a relationship which economists refer to as the *elasticity of demand*. A firm needs to have a reasonable idea of the actual relationship between price and quantity so that the effect on the firm's revenue of a price change can be assessed.

The relationship between costs and revenue

It is important to take an overall view of the profit pyramid. To simply observe that profits are at a satisfactory level is not enough. For example, a high profit level in a given period may be entirely due to a temporary market advantage which allows the firm to charge a very high selling price for its products. However, this may conceal problems of low productivity, and costs that are higher than they need be. This may become apparent later, and result in a poor profit performance once the temporary market advantage has been eroded by, say, new competitors.

Although revenue and costs have been shown to have a measure of independence from each other there is, for the firm, also an important relationship. A firm cannot simply decide, at least not in the short term, which quantity sold will generate maximum revenue and simply adjust production levels accordingly. This is because there may be an effect on costs. If a factory has an optimum capacity which minimises costs, then increasing production above this capacity could result in increased costs due to factors such as the need for overtime working and machine breakdowns. This might outweigh any gains in revenue. In the longer term, however, a firm may adjust its capacity to any level it wishes, by adjusting the level of fixed resources employed.

Conflict between costs and revenue — the impact of economic conditions

It has been shown that to improve profit performance, a firm must seek to increase revenue and reduce costs. It should, however, be questioned whether the conditions for satisfying these two requirements are likely to be present simultaneously. It is helpful to consider the position, bearing in mind the economic circumstances.

In times of 'good' economic conditions — that is, growing output and full employment — it is relatively easy for firms to sell their products because more people have money to spend. However, this may not be the best time for controlling costs because high levels of employment improves the bargaining power of labour. This may lead to higher wages and patterns of working conditions which do not maximise productivity, thus resulting in higher costs. However, this tended to be less of a focus for concern during the long post-war boom, because firms were selling their products and making substantial profits.

In times of 'bad' economic conditions — that is, static or falling output and high unemployment — the position is reversed. Management is in a position to control costs more readily by limiting wages and implementing higher productivity methods of working. However, the problem becomes that firms find difficulty in selling their products because there is less spending power and/or low consumer confidence. This has become a focus of more obvious concern in the recessions of the 1980s and 1990s.

This apparent contradiction has a micro and a macro element. From the micro perspective, it is always possible that individual firms might be managed in such a way that they can overcome the problem and still thrive, even in bad economic conditions. For example, a firm may acquire a secure market position through innovative design, which enables it to continue generating high revenue. Another approach might be to attempt to break into new markets, possibly overseas.

From a macro perspective, it is more difficult to see how every firm could achieve success, and the overall outcome is still likely to be recession. It is mainly because of this line of reasoning that a radical perspective of management, as described in Chapter 2, tends to argue that the market system, where firms are mainly driven by the need to make a profit, will sooner or later find itself in crisis due to this inherent conflict within firms. However, many of the other schools of thought tend to play down this element of conflict and argue that each organisation has the ability to manage itself out of the recession. The question of conflicting objectives within organisations will be considered in greater detail in Chapter 5.

Liquidity

The need for profitability as a prerequisite for long-term survival has been established. However, it is still the case that profitable firms may fail to survive due to short-term difficulties such as a lack of liquidity. This could arise where a firm is unable to generate sufficient liquid funds, especially cash, to pay its bills. Hence, this is often referred to as a cash flow problem. A major reason why this occurs is that firms often rely on a bank overdraft to keep themselves afloat. If the overdraft facility is not extended, or is terminated, then survival is threatened.

The vulnerability of a firm to cash flow problems depends on a number of factors, including:

- its capital base
- its position in the payments chain.

Capital base

Where a firm relies on an overdraft, the continued support of the bank is necessary if the firm is to stay in business. The banks are often regarded as risk adverse, and inclined to withdraw support from firms at the first sign of trouble, in order to protect their own funds. The firms most likely to be affected in this way are those which have limited capital assets. This is because banks tend to demand collateral, or security, for advancing funds, and the firm with limited fixed assets will have little to offer. This problem affects some industries more than others. For example, a manufacturer might have factory buildings and machinery to offer as collateral, whereas a building contractor may have very little by comparison.

In addition, the position is exacerbated when lending policies become erratic. For example, at certain times, such as in the late 1980s, money has been relatively easy to obtain, while at other times it may be extremely difficult.

The payments chain

In many industries there is a lengthy and complicated sequence of commercial relationships. As a consequence, there are a whole series of payments which are normally subject to terms of credit. Indeed, relying on trade credit is a form of short-term finance which firms often use in addition to or even instead of bank overdrafts. However, problems arise because not all firms can rely equally on this. Much will depend on the actual terms of credit. If a firm can arrange matters so that those who owe it money, called *debtors,* pay in 20 days, whereas those to whom money is owed, called *creditors,* are paid in 30 days, then the firm will always be using someone else's money to finance its activities.

Another factor which influences the payments chain is the relative competitive strengths of the firms at each link of the chain. This is an issue which will be dealt with in more detail in Chapter 4. However, the main principle is that those with greater market power can impose more onerous terms on those with whom they deal, such as suppliers and customers, should they have lesser market power. In addition to imposing onerous terms there is also the suspicion that some large firms do not even abide by them, but instead delay payments to their creditors as long as possible. This means that creditor firms are put under great pressure and their survival may be threatened through no fault of their own. In theory, a wronged firm may have legal redress, but in practice this can be a lengthy, expensive business, and may lead to being labelled a 'troublemaker', with all that implies for future business.

In the construction industry, the payments chain is a well-established feature which will be referred to at various times in this text.

Before leaving the question of liquidity, it is worth looking ahead to market strategy, to be considered in Chapter 9. When firms are contemplating expansion and diversification, the desire to improve liquidity can be an influential factor. For example, a firm which diversifies into a retail sector, where customers normally pay with cash rather than credit, may benefit from an improved cash flow to the business as a whole.

Objectives of organisations in the construction industry

There are a wide variety of organisations in the industry. The majority are firms whose objectives are likely to be similar to any other commercial organisation. Therefore, it is to be expected that organisations such as contractors, subcontractors, builders' merchants and manufacturers of materials and components would have normal commercial objectives including the need to make a profit. However, it is worth considering separately two other identifiable groups of organisations in the industry, namely **clients** and **consultants**, to ascertain whether they conform to this same general pattern.

Clients

Although the client is often thought of as the individual or organisation who initiates a construction project, the position is more complex. Clients build for a variety of reasons, sometimes for their own use and sometimes with the intention of selling or renting out for someone else to use. Furthermore, 'the client' may not be a single organisation, indeed in many cases, especially on large commercial projects, the client role may be divided among several organisations.

When constructing a building for another organisation to occupy and use, the client is being a 'professional client'. In such circumstances, it can reasonably be assumed that the main objective is financial profit, as for any other firm in business. The business in this case is simply being a client. However, when a client builds for owner occupation then it is likely to be for some user benefit, such as increased productivity for a manufacturer, or a social benefit such as in projects undertaken by local authorities.

The role of the client can be analysed by subdividing it into

- identifying and implementing the need
- assembling the land
- financing the project
- managing the building on completion.

For many projects, a single client body will embrace all these roles. However, in large commercial projects, it is not uncommon for the entrepreneurial role of identification and implementation to be carried out by a property developer; for local authorities to have a role in land assembly; and for a large financial institution to act as long-term financier.

Many client bodies are not necessarily commercial, especially in the public sector. Furthermore, although a client may be a commercial organisation itself, construction may form only a small part of its activities. In such circumstances, the construction part of its activities may not appear to be carried out along strict commercial lines. For example, an organisation commissioning a new head office may be looking for prestige and a high-quality 'no expense spared' building. Despite this, the wider objectives of the organisation as a whole will not be far away.

Consultants

There are a wide variety of consultant organisations — architects, surveyors, engineers, and so on. In terms of objectives, consultant practices have traditionally claimed to be 'professional' rather than commercial, and have therefore sought the maintenance of high standards above profit. To this end, professional institutions have

- laid down codes of professional conduct
- controlled entry to the profession through an examination and training structure leading to professional qualifications
- set fee scales which clients have to pay to obtain the services of consultants, thus ensuring a full professional service without cutting corners due to competition from other consultants

There are two contrasting ways of looking at this:

1. Controlling entry and charging a non-competitive fee ensures that the work carried out by professionals can be relied upon.
2. Controlling entry has the effect of restricting numbers, thus maintaining the rarity value and hence the income and status of professionals. Moreover, the fee scales are a kind of monopolistic pricing mechanism which gives consultants no incentive to offer value for money. In short, some of the professional institutions have similar characteristics to craft trade unions.

In recent years, it would seem that clients have been tending towards the second view. This is evident from the widespread insistence that consultants now usually have to compete for their work through fee bids. Part of the reason for this is that clients have become more cost-conscious and demanding as an increasing proportion of buildings are commissioned by professional clients who regard their buildings as profitable assets to sell or lease. Another reason is that consultant organisations themselves have become more commercial in their outlook, while insisting, of course, that professional standards can still be maintained. Some of the professional institutions have recognised these trends by abandoning fee scales and by allowing their members to form limited liability companies rather than partnerships which carry unlimited liability. This has set in motion a series of mergers and diversifications, reflecting changes in market structure, similar to the changes that have already occurred in contracting.

As a result of these changes, it is becoming more difficult to argue that 'professional' consultants have a different set of objectives to 'commercial' contractors. Many consultant organisations have clearly revelled in their new commercial freedoms. At the same time, the trend among contractors has been to emphasise their professionalism. This has been exemplified by the growing importance of procurement methods, such as design and build, which have given an enhanced role and greater responsibility to the professional builder. Another indicator of this change has been the successful launch of the Chartered Building Company scheme of the Chartered Institute of Building.

Summary

This chapter has considered the question of organisational objectives. The theme of profit has run throughout, but it has also been important to consider the more fundamental question of organisational survival. The problem with this, however, is that many of the survival functions are difficult to measure, with the exception of the financial variable of profit. Apart from profit, it has been essential to identify the other main financial objective of liquidity. It was shown that a lack of liquidity is just as likely to threaten the survival of the firm as a lack of profitability, a problem very familiar in the construction industry. Finally, consideration was given to whether organisations in the construction industry differ from the general pattern. Some of the issues introduced in this chapter will be further explored in the next two chapters.

Further reading

For an economic interpretation of organisational objectives, see Lavender (1990), ch. 4 (firms in general), application F (objectives of construction clients). For a classic view on survival, see Drucker (1972), ch. 9. Additional references: Cole (1993), ch. 19; Dixon (1991), ch. 3; Handy (1993), ch. 7; Megginson *et al.* (1992), pp. 174–9; Galbraith (1972).

4 External influences on organisations

Introduction

The two main sets of external influences on a commercial organisation are market pressures and public policy. These two sets of factors are not mutually exclusive, and indeed an important aspect of public policy is deciding how big a role the market should have. For example, since 1979 a core aspect of government policy has been that the market should play the most substantial role possible. The belief is that organisations will be at their best when subject to market disciplines.

It is important to understand, therefore, how markets behave in theory and practice, and the extent to which organisations are affected by them. In particular, consideration will be given to labour markets, capital markets, and market relationships in the construction project payments chain.

Of course markets are meant to work in the interest of consumers and society as a whole. Imperfections or failures in the market are one reason for the intervention of public policy. An overview of the effects of public policy will be given, including general policies, and those which affect organisations more specifically. Finally, brief consideration will be given to legal regulation through company law.

Market behaviour in theory

Conventional neo-classical market theory is underpinned by two major assumptions:

- the tendency to equilibrium and
- the perfectly competitive model.

The tendency to equilibrium

The disparate plans of buyers ('demand') and sellers ('supply') will be reconciled at an equilibrium price and quantity which balances the market. Thus there is no wasted production and everyone is satisfied. The mechanism for bringing the market into balance has been variously depicted by such concepts as the 'invisible hand', and later by a notional 'auctioneer'. If the equilibrium is disturbed, then the market has mechanisms for restoring balance. Thus:

1. In the case of a minor disturbance, such as a temporary shortage or glut, there is flexibility of price around the equilibrium.
2. In the case of a major disturbance, such as a change in taste or fashion which increases demand for, say, timber-framed at the expense of traditionally constructed housing, there is a reallocation of resources from one to the other, and a new equilibrium position is easily found.

The perfectly competitive model

This is an idealised vision of the way the market is meant to work in the best interests of society. If the conditions of perfect competition exist, then society benefits from:

- wider consumer choice and
- greater efficiency in the allocation and use of resources.

The actual conditions required can be summarised as:

- *many suppliers,* none of whom should have a dominant market position, thus giving consumers maximum choice
- *free entry* into the market so that competition can be maintained over a period of time
- *mobility of resources,* so that resources can easily be transferred from one activity to another in accordance with consumer demands
- *perfect knowledge* of the market by all participants (buyers and sellers), so that all can make informed decisions and behave rationally.

It is also possible to define perfect competition in terms of the price that would prevail in such conditions. Imagine a situation where relatively few firms are selling in a market with a fairly good profit margin. Assuming free entry to the market, additional firms will wish to compete for a slice of the profits. This will tend to compete prices downwards, a process which will continue until profit margins have been eliminated, and *price is equivalent to cost.* Once this point has been reached, there is no incentive for additional firms to enter the market. This process is illustrated in Figure 4.1.

There are a number of points emerging from this:

1. Cost, as defined here, must include an element of profit which allows the firm to remain intact through replacement of its depreciating assets. This is a similar concept to the 'survival minimum' discussed in Chapter 3.
2. By competing prices down to this level, only efficient, low-cost firms would survive, since any firm with a higher cost level and hence a higher price would lose all its sales to one of the many other firms offering a lower price.

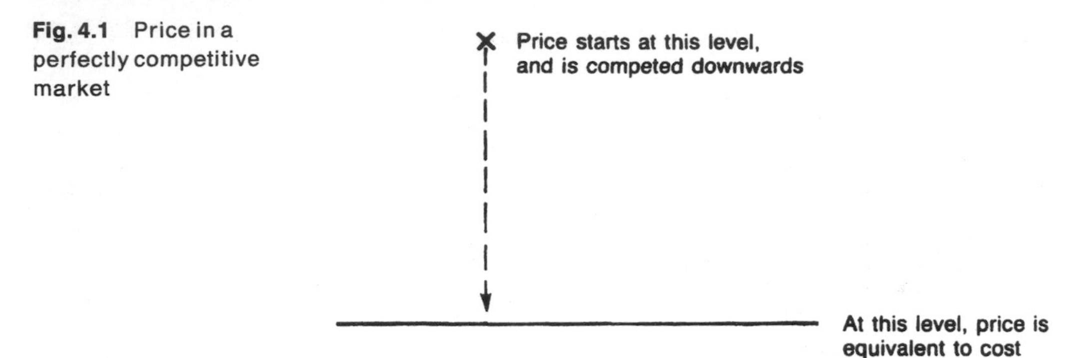

Fig. 4.1 Price in a perfectly competitive market

The significance of market theory to organisations

Few would argue that the assumptions of market theory hold exactly true in the real world. For example, it is hard to make a case that the housing market is often in equilibrium, or that all industries are highly competitive. However, many *would* argue that the benefits of something approaching perfect competition could be achieved if the barriers to the free workings of the market were removed. This view has been highly influential on public policy since 1979. Since then, the dominant economic philosophy has been monetarism, as described in Chapter 2. This will be more fully discussed later in this chapter, but one example will be given now. If the government believes that there are a good deal of inefficiencies, due to obstacles in the free market, then a tough monetary policy such as high interest rates may be implemented to 'squeeze out' these inefficiencies. This will pressurise the profitability and liquidity of firms and many may fail to survive.

Many would argue that perfect competition cannot be achieved in any circumstances, and it is pointless to try because in practice the market operates with many imperfections. This view will be considered next.

Market behaviour in practice

If the perfectly competitive model is not a true representation of the market behaviour of organisations, then it is necessary to consider which 'imperfections' might be present in the market, and how it might be possible to define the extent of competition in a real market.

Market imperfections

The conditions for perfect competition were stated above, that is, many suppliers, free entry, mobility of resources, perfect knowledge. In practice each of these conditions is likely to be violated.

Monopoly power

In many industries the market appears to be dominated by relatively few firms.

This could result in reduced choice for consumers, inefficiencies and higher prices. This situation is not only a disadvantage for consumers, but also for many firms because a substantial number of market transactions are conducted between two firms rather than between a firm and a consumer. Therefore, monopoly power can exist between firms, say, at different stages of the production process. For example, in the construction industry, monopoly power may be an issue not just between clients and contractors, but also between main contractors and subcontractors.

However, because there are few firms dominating a market this does not of itself result in monopolistic behaviour. As will be seen later, the 'degree of monopoly' in a market depends on additional factors, such as the strength of demand.

Barriers to entry

Free entry to a market may be hindered in a number of ways. Some of these may appear to arise quite naturally — for example, in the case of public utilities or other 'natural monopolies'. Other cases may be where large-scale production is required for economy. This situation means that economies of scale are present, usually requiring a large investment to set up, thus limiting the number of firms who are able to compete. In such situations competition would possibly be wasteful and some kind of monopoly power may be justifiable in the public interest.

However there are other less justifiable barriers which may restrict competition. These include:

1. Making potential competitors incur high advertising costs and face a multitude of brand names, thus artificially increasing costs.
2. Predatory pricing, by cutting prices below cost in the short term, so that new firms cannot afford to compete.
3. Maintaining excess capacity so that the potential exists for the market to be flooded in the event of new competitors seeking to enter the market.

It should be noted that there are costs involved in engaging in these practices, and firms may not find it worthwhile to do so. Instead they may choose to compete on the usual grounds of efficiency.

Immobility of resources

Firms do not necessarily find it easy to transfer resources from one activity to another in response to changes in demand. This is particularly true for large, highly capitalised firms who have made a large investment in a particular business activity. Furthermore, all resources tend to exhibit characteristics of immobility.

1. Land is immobile in that it takes some time to change the use of a site, given the need for planning permission, demolition, and rebuilding.

2. Labour may be immobile in terms of inappropriate skills, in the wrong location.
3. Capital may be immobile in that long-term assets cannot easily be converted for an alternative use, or sold for their true value.

It may be possible for the firm to overcome some of these immobilities, for example by:

- planning well ahead to meet its land and building requirements
- training employees in new skills
- recruiting employees from elsewhere
- using flexible open plan factory and office space so that adjusting to different circumstances is easier
- leasing rather than buying, thereby tying up less capital (depending, of course, on the terms of the lease).

Other immobilities may be addressed through public policy, including:

- streamlined planning procedures
- publicly financed training schemes
- investment grants and similar incentives to UK or overseas firms.

Imperfect knowledge

The concept of perfect knowledge embraces rational behaviour by all participants in the market. The problem is that consumers may not always make the right decision when confronted by a huge array of choice, and a lack of technical knowledge about the products. Of course, firms may exploit this in their marketing, through product differentiation.

Firms themselves do not always behave rationally, good examples being the lending patterns of banks and building societies in the 1980s along with the diversification policies of many building contractors into activities such as property development. It is increasingly recognised that the behaviour of consumers and producers is very likely to be influenced by the level of confidence — the so-called 'feel-good factor', or lack of it.

Defining competition in a market

Given that perfect competition does not exist, it is useful to have the means of assessing an actual market situation. A starting point is to focus on the price. It was shown earlier in this chapter that under perfect competition price would be competed down until it was equivalent to costs. Therefore it is logical to measure competitiveness in an actual market by the extent to which price deviates from the perfectly competitive price, that is, the mark-up of price over costs. This measure can be called the *degree of monopoly*, and by definition this will be zero under conditions of perfect competition.

A practical example, familiar in the construction industry, is the process of preparing a competitive tender bid. This is usually in two stages:

- estimating, where the costs to the firm of undertaking the project are calculated
- tendering, where the mark-up to be added to the estimate is decided.

In essence the process of estimating is calculating what the price would be under perfect competition, whereas tendering, by deciding on mark-up, is in effect assessing the degree of monopoly which the firm possesses in the market.

The degree of monopoly will depend on both supply and demand factors.

Supply

The most important factor is the number of firms competing in the market. Generally speaking, the fewer the firms, the higher the degree of monopoly. Since this is a structural factor, supply tends to remain stable over a reasonable period. Therefore it might be expected that mark-up will also remain stable, but this disregards the greater volatility of demand.

Demand

The strength of demand will also affect the degree of monopoly in that a strong demand will enable firms to increase mark-up, and vice versa. The strength of demand is best measured by *elasticity*, that is, the extent by which demand changes in response to a price change. Unlike the number of firms, the strength of demand is more liable to change, and tends to be rather more cyclical than structural. Thus, in times of high activity and strong demand, an industry dominated by a few firms will probably show signs of a high degree of monopoly. However, in difficult economic circumstances the same few firms will be compelled to compete with each other to a greater extent, with cuts in mark-ups more likely.

Another factor influencing the degree of monopoly is whether there is any **collusion** between firms to keep mark-up high. This, of course, is very hard to identify not least because of its dubious legality. However, when the other two factors are favourable — that is, few firms and strong demand — then the conditions may favour collusive behaviour.

The labour market

The labour market has great significance for both the firm and the economy as a whole. Indeed the condition of the labour market, particularly the level of unemployment, is often taken as the benchmark for the condition of the economy as a whole. This is due to its effect on consumer confidence and demand. For the individual firm the labour market affects wages and productivity, which in turn affect costs, as shown in the previous chapter.

A major significance of the labour market, especially the level of unemployment, is that it affects the balance of bargaining power between management and employees. This in turn affects wage levels, patterns of working practices and productivity performance. It also affects the extent of collective bargaining — low unemployment tends to extend its incidence, whereas high unemployment tends to reduce it. This issue will arise again in subsequent chapters.

The capital market

It was indicated in the previous chapter that the owners of a commercial organisation — that is, the shareholders — may not regard the firm's survival as paramount. The reason for this is that shares of publicly limited companies are traded on the stock market which forms part of the wider capital market. The workings of these markets can have a profound effect on the workings of an organisation.

The traditional vision of a shareholder is someone who is willing to share commercial risks on a long-term basis. In other words, shareholders expect that, over time, the company will succeed and the value of their shareholding will rise, resulting in a capital gain. There is also the expectation that income, or dividend, will be paid. However, this is not normally expected for some time, thus initially allowing all profits to be ploughed back into the business for expansion. Even at a later stage, it is the long-term investment which is more important than short-term income. This characteristic distinguishes owners' or shareholders' funds from borrowings. In the latter case, regular interest payments must be made which may restrict investment funds.

If the growth of a firm is tracked, a typical pattern might emerge:

1. The entrepreneur starts a private company by creating shares using a certain amount of owned or borrowed capital.
2. Over time, the firm will grow due to profits being made and ploughed back into the business, supplemented perhaps by modest borrowings.
3. At some point, the entrepreneur may wish to make a significant expansion, say through a major capital investment.

In the UK, this significant expansion has typically been accomplished by 'going public'. This entails offering shares in the company for sale to investors. Assuming the company has grown since the original shares were created, they should be worth considerably more than their face value. The effect is that the company receives an injection of capital by new investors coming in to share the risks with the entrepreneur on a long-term basis. At some point some shareholders may wish to sell their investment for a capital gain, but this is likely to be to an investor with similar long-term objectives. Thus the company can be sure of long-term funds for investment, unlike in other countries where there is greater dependence on bank borrowing with all that this entails for the burden of regular interest payments.

This idealised view of how the stock market is meant to work is, however, open to question. Apart from the 'traditional' shareholder, there are two other types that can be identified:

1. Those investors seeking a quick capital gain by buying and selling blocks of shares to take advantage of shifts in the market.
2. Those institutional investors, such as pension funds and insurance companies, who are primarily seeking a guaranteed income or dividend payment on a regular basis.

Many of these institutional investors will have a portfolio of investments, of which shareholdings are only one. In this electronic age the shifting of funds globally between investments is relatively easy and so the tendency for long-term investment may be diminished.

There are several consequences of this:

1. There may be insufficient funds for long-term investment because shareholders require a regular dividend irrespective of the performance of the company. This means that in terms of the financial management of the company, shareholders' funds offer no advantage over borrowings due to similar regular debt burdens.
2. If the shareholders' requirements are not met, be it quick capital gains or guaranteed dividends, then the shares may easily be sold, thus depressing share prices. This might make the firm vulnerable to a takeover bid because its stock market value may fall below the real asset value.

The question then arises: Which of the three types of shareholder described above predominates in the capital markets? It is estimated by Thomas (1994) that about half of all UK shares are owned by pension funds. While it cannot be assumed that all will behave in a short-term manner, the tendency may exist, because pension funds, in turn, have a regular commitment to pay incomes to their pensioners.

The workings of the capital markets described above also have implications for internal relationships in organisations, and this will be examined in the next chapter.

Market relationships in the construction industry

In Chapter 3, the concept of the payments chain, whereby firms trade at each link or stage in the production process, was introduced. Each link represents a market relationship or economic bargain which is regulated by a contract. In the construction process, the key relationship is between client and contractor. However, this may be extended since the contractor is likely to contract with subcontractors, who in turn may contract with subsubcontractors, etc. At the other end of the chain, the client may not be the ultimate occupier of the building, in which case there may be a contract with a tenant or purchaser. In addition, clients may contract with consultants, and contractors with suppliers. The whole area of contractual relationships in construction has been widely debated with a recent major contribution from the Latham Report (1994). Aspects of this debate will be considered in Chapter 24.

As discussed in Chapter 3, a major implication of the payments chain is that relative market strength at each stage will determine terms of the contract such as price and arrangements for making stage payments. These, in turn, affect organisational objectives such as profitability and liquidity. In common with many industries, construction seems to have evolved a structure whereby a relatively small number of large firms have a good deal of market power, while a large number of small firms are frequently in a subservient market position, often acting as subcontractors. This has often put great pressure on subcontractors, even threatening their survival.

The payments chain is one of the most significant features of the construction process, and is likely to feature strongly as an influential factor in any study of management in construction.

An overview of the impact of public policy on organisations

A major aspect of public policy is the extent to which intervention to supplement or replace the market should occur. This is very much a matter of perspective, as discussed in Chapter 2. The monetarists would argue that problems in the market occur because of imperfections; therefore public policy should be directed towards controlling inflation, and removing imperfections in goods, capital and labour markets so that the market can reassert itself. Critics of monetarism argue that the problems lie with the market itself. The so-called imperfections are really deep-seated failures which need correction through public intervention. Furthermore, if left to itself, the free market is likely to lead to a situation of severe cyclical fluctuations with periods of high unemployment.

The effect of this on organisations can be considered by reference to the conduct of economic policy since the 1940s. It can be seen that differences have occurred in a number of key areas such as public spending, taxation, interest rates, exchange rates, industrial policy and so on. A detailed discussion is outside the scope of this book but some main points will be considered.

There have been two main periods since the 1940s, each characterised by a different philosophy towards public policy.

1. *From 1945 till 1979* there was broadly a Keynesian demand management policy, necessary because it was considered that the free market was defective. Keynesianism involved pursuit of the main objective of full employment, with a tendency towards greater equality. This was to be

achieved by high levels of public spending to keep demand high, and by generally giving support to industry.

2. *Since 1979* the philosophy has changed to monetarism. This involves a belief that the free market is best. The main policy has been to eliminate inflation, this being a prerequisite for the market to work properly. At the same time, measures have been taken in the goods, capital and labour markets to remove perceived imperfections so that the market can be allowed to move as close as possible to perfect competition.

In assessing the effect of these policies, it is often argued that one effect of Keynesian demand management was that insufficient attention was paid to supply, with the result that it was not possible for the economy to adequately cope with increases in demand. The twin consequences of this were increased levels of imports, and inflationary pressure. Monetarism was meant to correct this by controlling inflation and rooting out inefficiencies through strict monetary controls, reductions in public expenditure and deregulation. Certainly the early 1980s witnessed a large number of closures and redundancies, supposedly the elimination of waste to make the economy 'leaner and fitter'. However, it must be questioned whether putting pressure on the company sector as a whole through a tight monetary regime really does root out the inefficient rather than simply allow the financially strong to survive at the expense of the financially weak. As noted in an earlier part of this chapter, it can be seen that in a payments chain the larger monopolistic firms can pressurise the smaller competitive firms. Thus, in a tight economic situation, the smaller firms may be more likely to fail than the larger, irrespective of levels of efficiency.

One test of monetarist policies is to observe what happens to the economy when a subsequent boom in demand takes place. This occurred in the late 1980s, and it appeared that the old problems of high imports and inflation, especially in housing and property markets, were once again with us. This would appear to imply that whether firms had become more efficient or not in the 1980s, the underlying weaknesses of the UK economy were still present. The productivity performance of UK industry will be considered in the context of production management in Chapter 10.

This overview of the impact of public policy has mainly discussed the general effects, but it must not be forgotten that there are numerous policies which may affect specific industries. Thus organisations in the construction industry are not only affected by general policies on interest rates and exchange rates, but also by specific policies on matters such as levels of public expenditure on housing and infrastructure, planning policies, health and safety policies, and many more.

Finally, the international aspect must be noted. Capital and goods markets are becoming increasingly global in character, with very large transnational companies in strong positions. Government policy on exchange rates can have an important impact. Since the Pound Sterling is no longer in the European exchange rate mechanism, its exchange value has decreased, making it easier for

UK firms to export. This brought something of an upturn in the economy, but this does not necessarily bring with it a boom in consumption, often the main engine of inflation. Whether an upturn is sustainable without inflation is the real test of whether changes in the economy and the management of organisations has been successful.

Legal regulation of organisations

The law, of course, is one aspect of public policy, and organisations are affected by it in many ways. This topic is generally outside the scope of this book, however employment law will be considered in Chapter 18. A very brief summary of company law will be given here, but the reader is recommended to refer to the section on further reading at the end of the chapter.

Commercial organisations may be partnerships or companies. Although consultants often form partnerships, most organisations in the construction industry are companies, which may be private or public (PLCs). Companies are distinct legal entities as compared to partnerships, which are a collection of fully accountable individuals. Companies have limited liability whereas partners are liable for their entire personal wealth.

Public companies differ from private ones, as their shares are traded on the stock market, whereas private companies restrict transferability. There are also more stringent requirements for public companies in respect of the publication of accounts.

The management of companies legally rests with the shareholders who appoint a board of directors at the annual general meeting to safeguard their interests. The directors in turn delegate to employed managers. In theory, shareholders may remove unsatisfactory directors, but in practice this may be difficult if directors are shareholders themselves, or if they have secure contracts of service which may be expensive to terminate. This returns to a point previously raised in this chapter, which questioned the interest of some shareholders in the long-term management of the company.

Summary

This chapter has considered some of the external influences on organisations. Particular attention was paid to the organisation's market position. The theory of market relationships was first considered, followed by an assessment of how the market works in practice. Of notable significance to the commercial organisation is its relationship to the labour and capital markets. For organisations in the construction process, the payments chain has a major impact on profitability and liquidity. Although market relationships are of great importance to organisations, the impact of public policy can also be extensive, not least because of its impact on market conditions. The general effect of public policy was considered together with some specific effects on particular industries such as construction. Finally, brief consideration was given to the legal regulation of organisations.

Further reading

This chapter is mainly concerned with external economic influences on the organisation. See the following for further discussion: Lavender (1990), ch. 3 (for markets in theory and practice) and ch. 7 (for public policy matters). For further information on legal regulation of organisations, see Galbraith and Stockdale (1993), ch. 4 or similar introductory legal textbook. There is a good deal of literature on contracts and their regulation of market relationships in construction. For a book with more of a management context, see Murdoch and Hughes (1992). See also articles from newspapers on current economic policy, such as Thomas (1994).

5 Internal influences within organisations

Introduction

It is all too easy to assume that the organisation is a coherent body with all its components pulling in the same direction. Indeed some of the management perspectives discussed in Chapter 2 tend to make this assumption. However, the commercial organisation consists of a wide range of people who may not all share the same objectives. For example, can it be readily assumed that all members of an organisation will take the same attitude to profit? Can it be assumed that profit maximisation is the common objective?

In this chapter, the various types of people in an organisation will be considered, to ascertain whether there is a divergence of objectives. A key factor is to consider whether there is a separation between ownership and control in an organisation, and whether it makes any difference if there is. There are a number of alternative theories of the firm that argue that there is a difference. Furthermore, it might be argued that there is inherent conflict between different levels of the management structure, and between management and other employees.

People in organisations

The firm is usually characterised, at least in conventional economic theory, as a coherent unified entity with all participants pulling in the same direction. Furthermore, it is usually assumed that the single overriding objective of the firm is to maximise profits, as discussed in Chapter 3. Firms come in all shapes and sizes, ranging from the individual operating as a sole trader, through to the massive multi-product transnational company. It may be easy to understand that the sole trader has a relatively free choice in deciding which objectives to pursue, but within the large organisation there may be scope for much debate. Can it be assumed that everyone within a large firm will be of like mind? Even in the context of a sole trader it cannot be assumed that profit maximisation will be the goal. For example, the skilled carpenter may prefer to engage in work which is most interesting and satisfying rather than most profitable. Similarly, an apprentice may be engaged because of a desire to 'pass on the trade' even though it might not be particularly cost-effective. The only constraint the sole

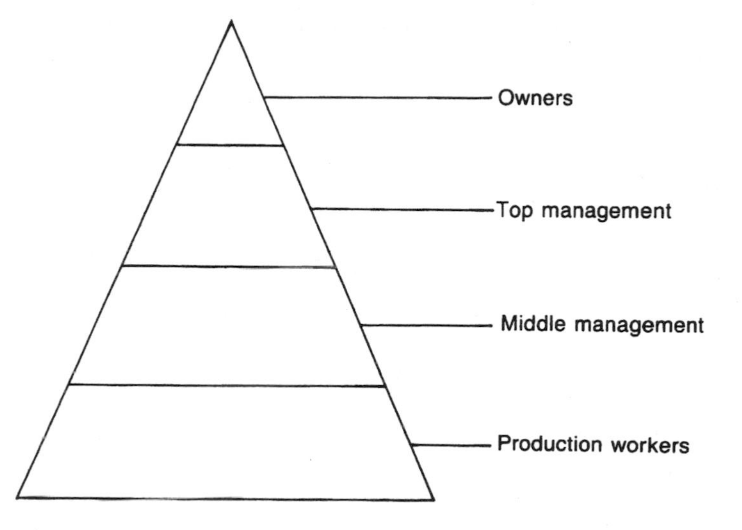

Fig. 5.1 Roles in firms

trader might feel is the need to make a satisfactory living. In some firms, especially where there are strong local connections, there may be a desire to create employment and contribute to the local community rather than to simply maximise profits. Again the constraint will be making a satisfactory profit to guarantee survival, as discussed in Chapter 3.

The above paragraph may give some indication of the diversity of motives possible, but does not assist a general understanding of organisational behaviour. To attempt this, it is necessary to identify the various roles which may occur within a firm, and consider whether any conflict of objectives is likely. For this purpose, four roles can be identified, as illustrated in Figure 5.1:

- owners
- top management
- middle management
- production workers.

Not all these roles will be separated in all firms. The degree of separation may depend largely on the size of the firm. For example, in the case of the sole trader, all roles will be combined; in the small local firm the owners may directly manage production. Generally speaking, the larger the firm, the greater will be the size of the middle management structure. As will be shown, it is at the middle management and production worker levels that most debate about possible conflict of objectives takes place.

- **Owners** Those who legally own the firm are partners or shareholders. Certain aspects of ownership were discussed in Chapter 4 — partners are liable for the entire debts of a partnership; companies may be private or public, with shares in public companies being tradable. As indicated above, owners may pursue objectives other than profit maximisation, but in the case of shareholders in public limited companies this is less likely.

- **Top management** This would normally be defined as the board of directors. In the terms of company law, as discussed in Chapter 4, the shareholders delegate the management of the company to the board of directors between annual general meetings. Since the directors are, theoretically at least, close to the shareholders, and may even be shareholders themselves, it is quite likely that they will share the same objectives.
- **Middle management** This includes the various managerial and supervisory employees below board level. The larger the organisation, the larger is the middle management structure likely to be. Similarly, in a large organisation, the top management tends to be more remote from the day-to-day management of the firm. It is principally for this reason that a good deal of debate ensues about whether true control of the firm rests with top or middle management. If it rests with middle management, then the concept of the **separation of ownership and control** comes into consideration, and the question arises of whether this separation affects the objectives of the organisation.
- **Production workers** This includes all manual and clerical employees who are involved in the firm's activities. The management perspectives, discussed in Chapter 2, reveal a variety of views on this issue, with some believing there is commonality of interest and others recognising inherent conflict.

These four roles are not entirely discrete. For example middle managers, and even production workers, may have shares in the company but, certainly in the case of the latter, this is unlikely to supersede their primary employee role. Similarly, some supervisors may spend part of their time in a production capacity, even if they are regarded primarily as part of the management structure.

Separation of ownership and control

This idea has been debated for some time but, according to Hillebrandt and Cannon (1989), the pioneer writers on the topic of separation of ownership and control were *Berle* and *Means*, writing in the 1930s. They had observed how manufacturing in the USA had increasingly become dominated by large firms where ownership was remote from management, thus affording greater powers to middle managers.

Identifying a separation is one thing, but showing that it makes any difference to organisational objectives is another matter entirely. A number of **alternative theories of the firm** have been developed which attempt to show objectives other than profit maximisation. Some of these will be considered, together with some criticisms.

Alternative theories of the firm

These theories are 'alternative' in the sense that they seek to depart from the normal assumption of profit maximisation. It has already been shown in Chapter 3 that many management writers concentrate on concepts such as survival needs. In this context there are two types of alternative theory:

- managerial
- behavioural

of which the former derive from economic theories, while the latter take a more multi-disciplinary view, incorporating sociology and psychology.

Managerial theories

These theories assume the separation of ownership and control, and that managers have the power to alter the objectives of the firm to suit their own needs. It is also assumed that the firm is not in a perfectly competitive market, thus allowing some discretion to price above costs.

There is still an element of maximising behaviour embedded in these theories, but instead of profit maximisation, some variable of management choice is maximised instead. Profit cannot be wholly ignored and it is recognised that sufficient profits must be made to satisfy shareholders and the survival needs of the organisation. Thus, whatever else may be maximised there is always a profit constraint. It will be apparent that there needs to be a significant degree of monopoly before managerial discretion becomes a realistic possibility.

The range of variables which may be maximised, instead of profit, include the following:

- sales revenue
- growth
- managerial utility,

each of which will be considered in turn.

Sales revenue

In this theory, managers attempt to maximise sales revenue, especially where high sales targets enhance the status of the managers. This could conflict with profit maximisation because profits are derived from costs as well as from revenue. Thus, if, in the quest for greater sales, production is increased above the most cost-effective level consistent with capacity, then marginal costs will increase faster than marginal revenue. This situation could occur due to increased overtime payments or more frequent machine breakdowns due to overuse.

Growth

Managers may feel that the best way to secure employment on high salaries and benefits into the future is by ensuring that the firm grows as large as possible. To achieve this, a good level of investment is required which may increase market share and profit in the long term, but which could easily depress profits and dividends in the short term. It might be expected that owners, too, would share this desire for long-term growth, but, as explained in the previous chapter, this may not be the view of many contemporary shareholders.

Managerial utility

This theory recognizes that managers may be interested in a range of benefits rather than one single benefit. Thus the firm will be managed with this objective in mind. The benefits which managers may seek include:

- high salaries
- fringe benefits
- number of subordinates
- power and discretion over the spending of budgets, especially on 'pet' projects.

Of course, profits and sales would also be important since these are likely to enhance status and, hence, the above benefits.

Behavioural theories

These depart from maximising economic models to consider a wider range of pressures on the organisation. There is no single objective which motivates the firm, but a whole raft of them, some of which conflict. Behavioural theories concentrate on the internal workings of the organisation, and on the decision-making process.

There are a number of features of behavioural theories:

1. It is not possible to concentrate on one variable to the exclusion of others, therefore the aim is to satisfy, or *satisfice* factors such as profit.
2. The organisation consists of a coalition of interests, not only owners or shareholders, but also employees, customers, creditors, the local community, the environment. These represent the stakeholders in the organisation.
3. To keep all members of the coalition happy it may be necessary to make 'side payments' which add to costs and are not profit maximising. This might include dividend payments or wage increases above what may be appropriate, given the firm's performance in the current year.
4. To keep every stakeholder happy there will need to be *organisational slack*, which is a buffer to help satisfy the various interests. It can only exist if there is a degree of monopoly, that is, mark-up of price over costs. Therefore slack cannot easily exist in a competitive situation.

Criticisms of managerialism

Although the concept of the separation of ownership and control pre-dates the

1940s, its adoption as a model for organisational behaviour dates from the post-war Keynesian/pluralist consensus as described in Chapter 2. This involved power shifting from shareholders to managers, who would then run the firm in the best interests of the whole coalition. This state of affairs, where the separation of ownership and control empowered managers to run the firm as they thought best, came to be known as **managerialism**. It was at its most influential in the 1960s and 1970s.

Even during this period there were criticisms. For example, economists such as Machlup (1967) were anxious to defend neo-classical theory. But from the radical social science literature there were critiques such as Blackburn (1972) which were later more widely accepted. It was argued that the 'managerial revolution' was a myth because the drive to make profits was still predominant, although it had to take a different form in the relatively full employment conditions of the time. Managers were seen to be still predominantly interested in profit because:

- they tended to share a similar background to owners, and to identify with them;
- capital market constraints prevented any alternative behaviour;
- organisations had been restructured to give financial control from the centre, preventing managerial autonomy.

Some of these points have already arisen, while others will be explored subsequently in this chapter and in later chapters. Before moving the managerial debate forward to the 1990s, some consideration will be given to the influence of production workers.

Production workers and their objectives

Some of the management perspectives discussed in Chapter 2 argue that there is no difference of interest between management and workers — what is good for the firm is good for the workers. However, it is also possible to argue that the objectives of production workers conflict in particular with the managerial objective of minimising costs — an important component of profit. This will obviously be the case in terms of wages — workers will want high income, which will add to costs. But this in itself may not be a problem for management. Indeed many managers will not seek to minimise wages if they see that paying good wages is the key to retaining better workers and thus improving performance.

Therefore it is the performance of production workers which is the key factor. This issue arose in Chapter 3 and will reappear at various points, for example in Chapter 10 on production management, and in Chapter 15 on motivation. As discussed in Chapter 3, performance is measured by productivity, and it must be asked whether workers will feel that striving to increase productivity is in their best interests. They may ask: What is the point in working hard for the benefit of shareholders and their profits? Managers may try to overcome this problem with financial incentives.

However, there is also the more fundamental question of the control of work. Skilled workers may wish to retain the right to plan and execute their own work, while non-skilled workers may seek to retain control through the control of shift patterns, staffing levels, and demarcation lines between tasks, all of which may make the work more amenable. In all these cases it might be claimed that 'the right of management to manage' is being questioned, and many industrial disputes are concerned with this sort of issue rather than with pay.

Production workers may have several outlets for seeking to achieve their objectives. Many will negotiate on an individual basis for better wages and conditions, particularly in white-collar jobs. It is also quite common for negotiations to take place on a collective basis, either officially through trade unions, or through informal groups. It is widely accepted that since 1979 there has been a swing away from collective bargaining towards more individual contracts. This has been encouraged by economic conditions in the labour market — that is, high unemployment — and through government policy on employment law. Chapters 17 and 18 will discuss this further.

Managerialism in the 1990s

Managerialism has undoubtedly been prominent in management literature, especially during the high tide of Keynesianism and pluralism in the 1960s and 1970s. However, the monetarist free market philosophy of the 1980s and 1990s would appear to have pushed these ideas into the background, and senior managers in particular are more likely to argue that their main task is 'to maximise benefits for shareholders'. This seems very much to be a restatement of the traditional view that the objective of the firm is to maximise profits. Thus the view of Blackburn (1972) that the 'managerial revolution' was a myth, or at least that it has not been sustainable, appears vindicated.

To assess the contemporary position, it is appropriate to discuss why management is more likely to pursue profit maximisation or 'maximum shareholder benefit' above all other objectives. The reasons fall into two broad categories:

- characteristics of managers
- external factors.

In certain cases the influence of production workers will have an effect.

Characteristics of managers

There are a good many reasons for arguing that managers have no fundamental difference with owners, and that the pursuit of profit will be uppermost in their minds:

1. The profit ethic is very strong. Even if the role of profit used to be played down, this has certainly not been the case since 1979. The philosophy of government has been to enhance the role of profit, giving greater legitimacy to profit-seeking behaviour.
2. The rewards paid to managers in terms of salaries and other benefits may closely depend on profit performance of the firm, with the ability to cut costs by 'downsizing' seeming to be particularly well rewarded.
3. A greater commonality of interest between owners and managers has been built through incentives such as profit-sharing schemes and share options.
4. Managers have increasingly sought to identify themselves with ownership by becoming owners themselves. One development in this respect has been the growth of 'management buy-outs', where a group of probably senior managers takes over the firm by buying the shares. This often happens with firms facing difficulties.

A further development has been the way in which managers have restructured firms to reflect the multi-divisional or M-form of organisation. In essence, this is a kind of devolved responsibility for carrying out tasks, but with strict financial control retained from the centre. This has been done to improve accountability within the whole organisation, and therefore acts as a control mechanism on both middle management and production workers. The full implications of this will be discussed in Chapter 7 on organisational structures.

External factors

At this point, it is necessary to refer back to Chapter 4, where external influences on organisations were discussed. It was shown that the two main sets of influences are:

- market pressures, and
- public policy.

The government's policy has been to increase the scope for market forces wherever possible, both in terms of influencing the private sector and by transferring major portions of the public sector to the private sector. This has amounted to a major structural change in the economy which has increased market pressures on firms. This increase in market competition has been accentuated by a parallel structural change referred to as the **globalisation of markets**. The world economy is heavily influenced by large companies operating on a transnational basis. Thus resources can be transferred between countries, a process aided by developments in information technology.

Two specific types of market explored in Chapter 4 were the labour market and the capital market. It was stated that the way in which they operate has implications for the internal workings of the organisation. For example, the state of the labour market will affect the relative bargaining power between managers and workers. The capital market, especially as operating through the takeover sanction, restricts managers in their pursuit of objectives other than profit maximisation. Globalisation has affected labour and capital as well as goods markets.

Apart from structural changes, market pressures have also increased in the 1990s due to cyclical conditions, especially the recession. The effect of this has been particularly noticeable in the labour market; where high levels of unemployment have greatly reduced the bargaining powers of employees.

The effect of the foregoing is that the degree of monopoly has diminished due to:

- structural factors, such as increased international competition
- cyclical factors, such as the weakness of demand caused by the recession.

This is far removed from the earlier period of the 1960s and 1970s when demand was stronger and international competition was probably less intense. In those times, higher degrees of monopoly provided a buffer to allow managerial discretion to an extent which does not exist now. This probably goes some way to explaining why managerialism is less, and the profit motive more, influential.

Summary

This chapter has considered internal influences on the organisation and particularly whether managers have the discretion to pursue objectives other than profit maximisation on behalf of the owners or shareholders. As a starting point, the diversity of people in an organisation was identified. It was noted that in a large organisation there is the possibility of a separation between ownership and control. This might make a difference to how the firm operates if the controlling managerial group are able to pursue alternative objectives. Alternative managerial and behavioural theories were then discussed together with criticisms from orthodox and radical literature. It has been argued that these theories have less support in the monetarist 1980s and 1990s than they had in the Keynesian 1960s and 1970s. The diminished degree of monopoly, due to increased global competition and the recession, has resulted in less discretion for managers to pursue objectives other than profit maximisation. The enhanced profit ethic of recent times makes it less likely that managers will wish to pursue alternative objectives in any event.

Further reading

For additional information on theories of the firm, see Lavender (1990), ch. 4 and

Hillebrandt and Cannon (1989), ch. 1. For criticisms of managerialism, see Blackburn (1972), ch. 8.

6 Financial structures

Introduction

Finance is an important theme of management. After all, the objectives of organisations are normally expressed in financial terms, as are many of the targets used to assess the performance of an organisation. Increasingly, senior positions in the management of organisations are occupied by those with financial management training and experience. Therefore it is to be expected that financial matters play a prominent part in a management textbook.

Financial management, therefore, has a prominent place in this book. Apart from any side references to financial topics, there are three chapters specifically dedicated to these matters:

1. This chapter considers basic financial images of firms — how they are financed, how the finance is used, how financial statements illuminate performance.
2. Chapter 11, which is within the management functions section of the book, considers how financial information is used to manage the activities of the firm — a process known as management accounting.
3. Chapter 22, within the management applied to construction section, considers financial control of construction projects from the point of view of the various participants.

In seeking to understand financial structures, it is most important to relate this to the objectives of the organisation as discussed in Chapter 3. The two main financial objectives were identified as:

1. Liquidity in the short term, meaning that cash flow or working capital is generated, thus enabling debts to be paid and the firm's current activities to be financed.
2. Profitability in the long term, meaning a surplus of revenues over costs, thus enabling the cycle of production to be renewed over time, and funds to be provided for replacing capital, and the expansion of activities.

As will be shown, financial management ultimately relates to achieving these objectives.

Sources of finance

All individuals and organisations need a variety of types of finance for different purposes. This is equally true for firms as it is for consumers, households and the government. For example, a family contemplating the purchase of a house is unlikely to finance this by a bank overdraft, but will instead turn to a specially designed source of finance for this purpose, namely a mortgage.

Similarly, firms need certain types of finance to provide for their short-term day-to-day needs of maintaining liquidity; and other types to provide funds for long-term investment which lays the foundations for profitability. Therefore, to summarise financial needs:

- short-term finance is required to meet the objective of liquidity
- long-term finance is required to meet the objective of profitability.

Each type of finance will be considered in turn.

Short-term finance

Short-term finance is needed to maintain liquidity, and is therefore the source of working capital to keep the firm operating on a daily basis. Many firms fail through lack of liquidity rather than lack of long-term profitability. For example:

1. A housebuilder may fail because customers have not yet parted with their money, rather than because the houses will not sell at a profitable price.
2. A builders' merchant may fail because contractors are slow to pay their invoices, rather than because the price of materials and quantity sold is inadequate for profitability.

Firms may use short-term finance in a number of ways. For example, they may:

- hold cash balances to pay immediate bills and for emergencies
- extend credit to customers — that is, debtors — possibly in accordance with standard practice in a given industry
- hold stocks of finished goods awaiting sale or delivery to customers
- hold stocks of materials and components awaiting incorporation into the production process
- pay for work in progress, in the case of ongoing work such as construction projects

The more working capital that is tied up in these items, the greater the strain on the firm's finances. Thus, an ineffective debt-collection mechanism, or the tendency to carry too high a level of stocks, means that more short-term finance has to be raised. This can be a particular burden at times of high interest rates, and can certainly threaten the firm's survival. It would seem that firms have become more aware of these problems in recent years. For example, a particularly significant change is the tendency to carry far fewer stocks at both ends of the production process — materials and components, and finished goods. This is partly due to greater caution in the recession, but also because of the influence of Japanese ideas such as 'just-in-time' which will be discussed under 'Production management' in Chapter 10.

Given that there will always be a need for some short-term finance, the main types available are:

- bank credit
- trade credit
- factoring.

Bank credit

The provision of short-term credit to firms, and to a lesser extent to consumers, has traditionally been the main function of commercial banks in the UK. This differs from banks in other European countries who have tended to see their role more in the realms of long-term finance. Indeed UK banks have been more inclined to follow this practice in recent years. Nevertheless, bank credit, especially the overdraft facility, represents the best-known method of short-term finance for firms in general.

Overdrafts work on the basis that a firm is granted certain drawing rights by the bank up to a prescribed limit. Thus the firm borrows on a flexible basis according to its needs at a given time. Therefore, the faster it is able to collect debts then the less it will need to borrow and, hence, the less interest charges will be incurred. So in a construction project, the quicker a contractor receives interim payments from the client the better.

Trade credit

This has always been a standard practice in construction and many other industries. Essentially it means that goods are supplied on a credit basis. A certain period, within which payment must be made, is normally prescribed, be it 14 days, 21 days, 28 days, etc. A firm using trade credit is therefore using somebody else's money to finance its activities. This means that it is using the other firm, or creditor, as if it were a bank. Using firms as creditors is the other side of the coin to extending credit to firms, as described above. Thus emerges the payments chain as described in Chapter 3.

The allowed credit period can vary greatly between industries and even between firms in the same industry. Indeed in some industries, especially retail, the credit period is zero and the customer normally pays in cash. It is worth reinforcing the point made when discussing the payments chain in Chapter 3, that firms sometimes exploit trade credit by delaying payments due to creditors as long as possible. This seems more likely where a situation of unequal market power exists, such as between large and small firms, and between main contractors and subcontractors.

Factoring

This type of credit involves the use of a third party to collect debts on a routine basis. Factors may be specialist financial organisations or subsidiaries of more general financial institutions. Where factoring is used, all invoices are issued through the factor, thus breaking a direct contact between supplier and customer. This is a possible disadvantage, although it should be noted that the supplier's name will still appear, possibly with some qualifying word or phrase in brackets — 'Bloggs (Sales)', instead of plain 'Bloggs'.

The way the system normally operates is that the factor will pay the firm a substantial proportion of the invoice value immediately, thus improving cash flow. The remainder is paid when the debt is recovered, less the fee payable to the factor. This may seem an expensive method of finance, but the firm benefits from swift payments. Also a valuable specialist administrative service is provided by the factor. This makes the system particularly appropriate for those firms who trade with large numbers of firms, with each transaction being of a relatively small value.

Long-term finance

Long-term finance is for investment in assets which will be used to enhance profitability in the long term. These assets may include:

- tangible items or fixed assets such as land, buildings, machinery and vehicles
- intangible assets such as goodwill, which may be built up over time.

These intangible items may be difficult to measure and value but certainly count as investment assets because finance would have been expended on them, perhaps in the form of employee training which has a long-term benefit for the effectiveness and reputation of the firm.

There are basically two main sources of long-term finance:

- owners' funds
- borrowed funds.

These will be considered separately, but a significant difference, from the point of view of financial management, is that borrowed funds normally carry a fixed and regular charge in the form of interest, while owners' funds, in theory at least, do not. It should be noted that in the UK there is a greater tradition of firms being financed through owners' funds than in many European countries where, as previously mentioned, the banks have had a more substantial role in the long-term financing of industry.

It should also be mentioned that this study of long-term finance refers to the general financing of the firm's investment expenditure. In the construction and property industries there are a number of specialist sources of finance for the creation of major property assets which will not be considered here.

Owners' funds
These include:

- funds provided by the original entrepreneur to start the company
- receipts from shares sold if the company 'goes public', a process described in Chapter 4
- any ploughed back profit retained in the business.

These funds represent the permanent capital of the firm. There are several types of shares or share issues, but it is not intended to discuss these here (see 'Further reading' at the end of the Chapter). It will be assumed that 'shares' refers to ordinary shares, or *equity* as it is often called.

In theory the owners, or shareholders, are risk-taking investors and will be prepared to forgo an income now in exchange for a greater return in the future. This should enable the firm to use the maximum amount of funds for investment and growth in the foreseeable period, without the burden of having to pay out regular interest charges. However, as pointed out in Chapter 4, modern shareholders may not conform to this behaviour. Instead they may insist on a regular income or dividend. Failure to provide this may lead to sales of shares, followed by a possible drop in share prices and the firm being taken over.

Borrowed funds
Funds may be borrowed from a number of sources, but share the characteristic that they represent a regular drain on the firm's financial resources due to interest charges. The most straightforward way to borrow is from banks or some other financial institution on a long-term basis. This can be for whatever agreed term is appropriate, and will carry a rate of interest. Alternatively the firm may issue debentures, also known as loan stock. Holders of these will also receive a fixed return, but debentures can also be bought and sold in the same way as shares.

Capital structure

Capital structure relates to the proportions of total capital which are, on the one hand, borrowed funds and, on the other hand, equity (owners') funds. This proportion is known as capital **gearing**. The higher the proportion of borrowed funds to owners' funds, that is, non-equity to equity, then the higher is said to be the gearing. The reason for making this distinction was previously mentioned — borrowed funds carry a definite cost against the business, whereas owners' funds do not, or at least are not meant to. In deciding what gearing strategy a firm should adopt, some additional points need to be considered:

1. Lenders will feel their investment is fairly secure if there are sufficient shareholders to carry the risk, therefore, the firm will be able to borrow at reasonable rates of interest.
2. However, as borrowings increase — that is, gearing becomes higher — lenders will feel less secure as the number of shareholders taking the risk diminishes. Therefore higher interest rates will be demanded as a risk premium.
3. Shareholders may be attracted by high gearing, because in times of high profit the lenders will still receive a fixed return, leaving a greater amount to be shared among the fewer equity shareholders.
4. However, shareholders in this highly geared situation are carrying a more concentrated risk, so they will require a higher potential return to attract them.
5. If the management of the firm wishes to undertake fast growth, then high borrowings may be the only way to raise the required funds quickly enough.

Choices have to be made, balancing all these factors. If there is a booming property market there may be a temptation to strive for fast growth by borrowing heavily. This carries the risk that if interest rates rise, the regular debt payments will soar, threatening the survival of the firm due to a lack of liquidity. This is an all too familiar story in the construction and property industries of the 1990s following the boom and subsequent bust of the late 1980s.

Financial statements

An enormous amount of financial information exists about a firm, and there are legal obligations that some of this information has to be published and made available for investors, customers and the public at large. Obligations to disclose information are greatest for public limited companies. The usual outlet for this is the report and accounts prepared for the annual general meeting. The following three financial statements are usually included:

- balance sheet
- profit and loss account
- sources and application of funds statement.

The first two are well established; the third is more recent and included largely in response to criticisms of the other two. It should be noted that managers will have considerably more financial information for internal use. The published accounts represent the external financial face of the firm. A brief consideration of the principles underlying the financial statements will follow. Further details and numerical examples can be found in many accountancy texts (see 'Further reading').

The balance sheet

The balance sheet is possibly the best-known statement about a firm's financial position. It is potentially also one of the most misleading. The reason for this is that the balance sheet does not measure performance over a period of time, but is more of an instantaneous 'snapshot'. Thus, a firm's balance sheet may be very different today from what it was yesterday, and from what it may be tomorrow. In the normal course of events this situation is unlikely, but the fact that it is possible makes the balance sheet more susceptible to manipulation, thus giving a false impression of the firm's financial position.

The balance sheet attempts to show what financial resources are available to the firm, and how they are being used. It must balance because more resources cannot be used than are available. Thus, the main vertical division on a balance sheet is between:

- where finance has come from — **liabilities**, and
- how finance is being used — **assets.**

There is also a horizontal division in the balance sheet, to reflect long- and short-term objectives, with short-term items usually described as *current*.

Therefore the balance sheet essentially consists of a matrix of four sectors, as shown in Figure 6.1. It has already been established that liabilities must equal assets. It is also highly desirable that there is broad parity between long-term liabilities and assets; and between short-term or current liabilities and assets. This would indicate that the appropriate type of finance was being used for the appropriate purpose. A brief discussion of each of the four sectors follows.

Long-term liabilities

These are essentially the sources of long-term finance previously discussed — that is, owners' funds and borrowed funds. All the different types are listed including

	LIABILITIES (where money comes from)	ASSETS (how money is used)
LONG TERM	Owners' funds Borrowed funds	Buildings Plant and machinery Vehicles (according to industry)
SHORT TERM (current)	Creditors Overdrafts	Stock Debtors Cash

Fig. 6.1 Sectors of the balance sheet

ordinary shares, preference shares, debentures, loans. Also included will be reserves, which are retained profits accumulated over the years.

Long-term assets

This shows the uses to which the long-term finance has been put. It will vary according to the type of industry in which the firm operates. For a manufacturer, it might include factory premises and machinery; for a building contractor it might include head office, plant and vehicles; for a property company it might include buildings being let and land banks. A problem associated with long-term assets in balance sheets is: What value should be ascribed to them? Should value be based on original or historic cost, or should some allowance be made for inflation, replacement cost, or speculative changes in value? This is a problem much debated among accountants. Different answers can give very different results.

Current liabilities

These are essentially the sources of short-term finance previously discussed, including trade creditors and bank overdrafts.

Current assets

Unlike a long-term asset, a current asset is not expected to last throughout the accounting period. All firms are likely to have some cash, in bank and/or in hand, at any one time. Also, most firms are likely to have debtors, representing assets to be realised in the foreseeable future. Beyond this there will again be differences according to industry. A manufacturer may have stocks of materials, components, and finished goods; a building contractor may have materials on site and work in progress; the retailer will have goods in the shop.

A final point about balance sheets is to reinforce the point that they can change from day to day. For example, the purchase, disposal or revaluation of an asset can make a significant difference. The balance sheet can therefore give a distorted picture of the firm's financial position over an accounting period.

The profit and loss account

Unlike the balance sheet, the profit and loss account measures the firm's financial position over a period of time. It is therefore sometimes felt to be more reliable, and less vulnerable to manipulation.

The profit and loss account may be thought of as structured around the simple equation introduced in Chapter 3, that is:

PROFIT = REVENUE – COSTS.

The largest amount of the three is revenue, or income. From this are deducted various categories of costs to arrive at a figure for profit.

One problem in drawing up a profit and loss account is that the end of the accounting period will probably leave some loose ends. For example:

- goods may have been ordered and delivered to customers but not yet paid for (an expected revenue)
- the market value of goods held in stock may be greater than price paid (an expected revenue)
- goods and services received from suppliers may not yet have been paid for (an expected cost).

The general principle of accounting practice is to be cautious. Hence, expected revenue will be excluded from the accounts, but expected costs will be included.

Another problem relates to the definition of costs. It was shown in Chapter 3 that resources can be variable or fixed. These categories are extended into defining costs:

1. Variable costs apply to resources wholly used up in production, such as materials and labour time.
2. Fixed costs apply to goods partly used up in production, such as plant and machinery.

It is easy to see that variable costs are charged against revenue. However, charging for fixed costs is more of a problem. The solution is to charge a proportion of the cost based on the expected life span of the asset. Therefore, a machine with an expected life span of 10 years will usually have 10 per cent of its cost charged each year. As mentioned in Chapter 3, this process of gradually writing off an asset over its life span is referred to as allowing for **depreciation.**

Thus the structure of a profit and loss account is typically as follows:

(Sales revenue) – (Cost of sales) = **Gross profit**

(Gross profit) – (Operating costs) = **Net profit**

The cost of sales is broadly speaking the variable costs, whereas the operating costs are broadly speaking fixed costs or overheads and will include such items as administration, rent, heating and marketing costs, as well as depreciation allowances.

The net profit will then be further divided into:

- corporation tax paid to the government
- dividends paid to shareholders
- profit retained in the firm for long-term funding.

As is the case with the balance sheet, the exact contents of each category will depend on the industry.

Sources and application of funds statement

Whereas the balance sheet considers a moment in time, and the profit and loss account a period of time, this statement is designed to show how the financial position of the firm has *changed over a period of time.* It is prepared by noting the position at the beginning and end of the accounting period, and by observing what happened during that period. Therefore, the information required is the opening and closing balance sheets, plus certain data from the profit and loss account.

From the point of view of management, the sources and application of funds statement can give a more helpful picture of the firm's finances. This is because it shows how funds have changed and is therefore a better guide to the resources at the disposal of management, particularly in terms of liquidity and cash flow. One of the main explanations for this is that some of the items in the other two statements contain notional figures which may not reflect actual movements of funds in or out of the firm. For example:

1. Assets shown in the balance sheet reflect historic costs and do not show recent purchases or sales of assets which can make a major difference to resources at the disposal of management.
2. Profits are adjusted in the profit and loss account to allow for depreciation, which is only a notional allowance, and does not represent a cash flow out of the firm.

By making appropriate adjustments and studying all the financial statements, it is easier for managers to ascertain not just whether a profit has been made, but where it has come from and what has happened to it.

As mentioned, the sources and application of funds statement is prepared mainly by reference to the opening and closing balance sheets. In comparing the two, **sources of funds** comprise:

- increases in liabilities
- decreases in assets

and **application of funds** comprises:

- decreases in liabilities
- increases in assets.

If applications exceed sources, this represents a deficit for the period, likely to be met by a reduction in the cash balance held at the bank. Conversely, if sources exceed applications, then this is likely to represent an increase in the cash balance. Examples of the items which could appear in the sources and application of funds statement follow.

Sources of funds

Increases in liabilities might include:

- increase in ordinary share capital
- new loans taken out
- increases in retained profit, ignoring depreciation (this information from profit and loss account)
- increase in creditors, meaning more use of funds of others.

Decreases in assets might include:

- sales of fixed assets such as land, buildings, machinery and vehicles
- reduction in debtors, meaning less use of the firm's funds by others
- lower stock levels.

Application of funds
Decreases in liabilities might include:

- repayment of loans
- reduction in retained profits
- reduction in creditors.

Increases in assets might include:

- purchases of fixed assets
- increase in debtors
- higher stock levels.

Summary

This chapter has examined the financial structures of organisations. In doing this, it is extremely important to relate financial structures to the financial objectives of the firm. Therefore, the two objectives of profitability and liquidity have played a prominent part in this chapter.

The main topics considered were the sources of finance available to help achieve objectives, and how the firm might best structure its capital requirements. This led to consideration of the major financial statements which assist in an understanding of the firm's financial position. These statements are required, by law, to be published, and a good deal of useful information can be drawn from them if carefully read.

In Part Two of this book ('Functions of management'), finance will be considered more closely from the point of view of managers. It has already been shown that useful management information can be drawn from, say, the sources and application of funds statement. This will be extended, in Chapter 11, to consider a wider range of information which can be drawn from the accounts, starting with financial ratios relating to matters such as profitability and liquidity.

Further reading

There are numerous books on accountancy in print, many of which are specifically designed for the non-specialist reader. An example of a suitable text for further reading, particularly for the purpose of studying numerical examples to supplement the principles examined in this chapter, is Davies (1992), in particular chs 1,2,3,6,14,15 and 16.

7 Organisational structures

Introduction

Management will seek to structure an organisation in order to best fulfil its objectives. It is necessary to classify activities and set in place reporting mechanisms to achieve these objectives. For the firm, this means organising so that decisions affecting profitability and liquidity can be taken correctly in the appropriate time span.

The first topic to consider will be the theoretical ideas which seek to explain organisational structures. One of the most influential of these is **bureaucracy**, as first explained by Weber. Next the different ways of grouping a firm's activities will be considered. This sometimes results in diagrammatical representations of structure, that is, organisation charts. Within each type of structure there are various matters of hierarchy and relationships to be considered such as 'flatness' of organisation, spans of control, extent of delegation and decentralisation.

After these points have been discussed, special consideration will be given to the multi-divisional or M-form organisation, since this has emerged as the dominant form of organisational structure in recent years.

Finally, consideration will be given to communications in organisations. These are the various reporting mechanisms put in place to make sure that instructions are given and tasks carried out. A good deal of this may be formalised, for example, through reports and committees. However, it should be remembered that alongside formal structures, informal structures and lines of communication will also exist.

Explaining organisational structures

It is possible to conceive of the very small firm having virtually no formal structure to speak of, with the owner working alongside a limited number of workers who are given instructions and directly supervised. In such a situation it is quite likely that the employees will be working as a team, all doing more or less the same thing. Examples might include small bricklaying or carpentry gangs. Beyond this size of firm, there is likely to be a need for a division of labour, and hence some kind of structure. Therefore, as the organisation grows there emerges a differentiation of task which tends to separate parts of the organisation. To prevent this happening there is a need to integrate tasks. This apparent simultaneous need for both *differentiation* and *integration* was identified by Lawrence and Lorsch in 1967, and has often been used to explain the tension in organisational structures.

A similar approach is suggested by Handy (1993), who states that 'organisational structures need to tread a tight-rope stretched between the pressures for uniformity on the one hand and diversity on the other'. In brief:

- *uniformity* includes ensuring certainty and control of the firm's activities and output, while
- *diversity* includes taking advantage of specialisation, and remaining flexible in the market.

In Chapter 2, systems theory was discussed as one of the management schools of thought. This is a multi-dimensional way of looking at an organisation as a whole, and as such it might be expected to say something about organisational structures. It does to the extent that systems have to be structured to take account of the various subsystems within. Furthermore, open systems are affected by the environment, and account must be taken of the degree of openness when designing structures. However, the branch of management thought which probably gives most practical insight into organisational structures is bureaucracy.

The meaning of bureaucracy

This word has, of course, a rather negative image because it conjures up impersonal and inefficient organisations. Few would claim to be bureaucratic or express a desire to work in a bureaucratic organisation. However, the truth is that large numbers of people, perhaps even a majority, are most comfortable in an organisation where they, as Handy (1993) puts it, 'know where they stand, what they have to do, who is in charge and what the rules are'.

Studies of bureaucracy stem from the work of the sociologist Weber. His main interest was in how **authority** is exercised in society and why people obeyed. The word authority is distinguishable from the word **power** in that:

- authority implies acceptance by those affected, but
- power implies unilateral force to compel acceptance.

As will be seen shortly, the difference is not always clear.

There are three types of authority as identified by Weber:

- traditional
- charismatic
- rational-legal.

Each will be considered in turn.

Traditional authority

This type of authority exists where acceptance of rules are based on some tradition, custom or practice. This may exist outside the world of work, particularly as exercised by, for example, religious leaders. This may mean acting on what you are told irrespective of personal respect for the person 'in authority'. Within the world of work, traditional authority can be effective in some organisations. Individuals may respond to authority if they have been brought up always to do so.

Charismatic authority

This type of authority depends very much on the person exercising it. Perhaps the person has certain magnetic qualities or other personality characteristics which inspire loyalty in general or a course of action in particular. Charismatic authority may exist in addition to the other types. For example, religious and political leaders may be listened to for traditional reasons, as above. However, they may be listened to more closely, or by more people, if they are also great orators. Similarly, in a firm, the position of manager may carry authority for traditional or rational-legal reasons, but this may be enhanced if the individual is charismatic.

Rational-legal authority

This type of authority is what one would expect to see in a firm, because it derives from the official position of a person. In principle, the authority continues whoever is occupying the position, although, as indicated above, the authority can be reinforced by charismatic authority. Conversely, there are instances when a manager, who is given authority, exercises it so poorly that organisational performance suffers, and the manager is removed from the position of authority. Inherent in rational-legal authority are likely to be rules or procedures which define it. To take some examples:

1. The limits of the authority of managers at different levels needs to be understood.
2. The authority of trade union officials to call for industrial action needs to be defined.

These rules and procedures may be laid down by a variety of means, including company policy, collective bargaining agreements or by law.

Comments and criticisms regarding bureaucracy

A distinction was previously made between authority and power. It would seem, however, that the difference between rational-legal authority and power can sometimes appear rather slight. For example, in times of high unemployment it is not uncommon for managers to cut wages, implement more onerous working practices and revoke collective bargaining rights. This could be interpreted as:

- the unilateral imposition of less favourable conditions on workers, who have no option but to tolerate them, as the alternative is loss of employment and little prospect of any alternative; or

- the legitimate changing of wages and conditions of employment in accordance with changing market conditions.

The first is *power*, while the second is *rational-legal authority*, and an objective answer to which of the two interpretations is correct may be difficult to establish.

Other common criticisms of bureaucracy revolve around its inherent rigidity, for example:

- there is less incentive to look for unusual solutions to a difficult problem, and indeed this may be discouraged;
- the rules and regulations become ends in themselves;
- excessive paperwork stifles creativity and slows down communications inside and outside the organisation;
- concentration on roles reduces the human element and de-personalises the organisation;
- where very clear roles are specified, persons holding these positions may only be interested in their own specialism, and fail to see the overall objectives of the organisation.

These criticisms are quite telling, but the problem remains of how to structure a large organisation. It might be argued that with good management the benefits of bureaucracy can be obtained while overcoming the problems.

More recent views on bureaucracy

It has been argued, by Gouldner (1955) that the pattern of bureaucracy favoured by Weber — that is, rational-legal — can exist in *three* forms even within the same organisation.

1. **Mock bureaucracy**, where rules and procedures imposed from outside the working unit, say from head office, would be ignored or merely paid lip service to, this being an example of informal structures overlaying the formal.
2. **Representative bureaucracy**, where there was agreement between managers and workers that the rules and procedures were sound.
3. **Punishment-centred bureaucracy**, where one side or the other imposes its rules and procedures on the other, unwilling, party.

The value of this framework is that it allows for more variation in human behaviour. It also emphasises the importance of informal structures in influencing organisational performance. This differs from Weber, where emphasis on rationality implies that there is a single best way of structuring an organisation. The idea of the punishment-centred bureaucracy reinforces the earlier point that the distinction between power and authority is not necessarily as clear-cut as Weber might imply.

In conclusion, the criticisms do not alter the fact that organisations of any size are likely to have some measure of bureaucracy in their structure. The task of management is to obtain the benefits while minimising the disadvantages.

Grouping organisational activities

The larger the firm, the more difficult it might be to group the various activities in a way that makes sense and is easy to manage. In this part of the chapter, the various basic ways of grouping activities will be discussed separately. In reality, a combination of methods may be used to create a multi-divisional structure. Since this is the most common contemporary approach, it will be considered separately later. In whichever form activities are grouped, it is usually possible to show the result diagrammatically with an organisational chart. There are several different ways of grouping activities, which will be briefly explained.

- **By function** This is the traditional method of grouping activities. The firm is divided into departments, each of which is responsible for functions such as:

  — production
  — marketing
  — finance
  — personnel.

  These departments would tend to relate to the whole firm. There is a connection with achieving the firm's objectives in that the production department incurs costs and the marketing department generates revenue. However, these connections are limited compared to the multi-divisional structure.
- **By product** As a firm expands it may become difficult to organise on a functional departmental basis. For example, a single production department may be unable to deal effectively with several products, particularly where production is organised on a multi-plant basis.
- **By geography** A firm's activities may be spread out geographically. Typically, this may happen where a firm has grown through acquisition and merger and finds itself with production units located in various parts of the country. For the multi-national company, facilities may be located throughout the world. In these cases a geographical organisation may be the most effective structure. Another reason for a geographical structure could be for marketing purposes. For example, a structure which emphasises the proximity to particular markets will create a stronger regional identity for the firm's products.
- **By customer** This may be appropriate where a firm has a number of well-defined customers, or groups of customers, which define its work. Examples

might include public sector clients, commercial clients and so on. This is quite common in construction organisations. For example, firms of architects or surveyors may be organised into teams, each of which predominantly deals with a particular client on a regular basis.

- **By capital** It may sometimes be the case that a firm's activities centre around particular major items of capital such as buildings, plant or equipment. Examples might include:

  — major building developments owned and managed by a property company
  — production lines in a factory
  — dry docks in a shipyard
  — tower cranes owned by a plant hire company.

  These major items of capital are the main cost centres of the firm, and the bases on which revenue is generated.

- **By project** For some firms, activities may be defined by a relatively small number of discrete projects. In such cases the firm may be structured around project teams with limited head office support. As is the case with grouping by capital described above, the project is the main cost centre and basis for generating revenue. A good example of this type of structure is the traditional building contractor where the workload is entirely made up of a limited number of projects. Of course the majority of construction firms do not restrict themselves to contracting these days, but may also be involved in activities such as speculative housebuilding and property development.

These methods of grouping the firm's activities do not necessarily exist in isolation. Real firms are likely to have a structure which combines different methods of grouping, probably within a multi-divisional framework. Before examining this, some other principles of structure will be considered.

Hierarchies and relationships

Under this general heading there are a number of interrelated topics, as follows:

- depth of hierarchy
- span of control
- delegation
- centralisation/decentralisation.

These will be given separate consideration.

Depth of hierarchy This is concerned with the 'flatness' of the organisational structure, that is, the number of levels between top and bottom. Organisations can be flat or tall, with the choice depending on factors such as:

- size of firm
- technology involved
- management preferences.

Whatever the outcome, flat organisations tend to have fewer levels and relatively wide spans of control, whereas tall organisations tend to have more levels and relatively narrow spans of control.

Size of firm
Generally speaking the smaller the organisation, the fewer the number of levels. As size increases, the number of levels tends to increase, although this might be avoided by increasing the span of control.

Technology involved
The well-known studies of different manufacturing methods, by Joan Woodward, found that technology strongly influenced levels and spans of control. Categories of methods of production are:

- unit
- batch
- mass
- process

These will be discussed in 'Production management' in Chapter 10. In general, hierarchies were found to be taller and narrower in process compared to unit production, but spans of control were greatest in mass production.

Management preferences
In some firms, top management has a preference for devolving authority to middle managers, while in other firms there will be greater centralisation, and less autonomy at lower levels. This could affect choice of structure, but there are ambiguities, depending on how far down the line authority is to be devolved. For example:

1. A *flat structure* allows devolved authority to middle management, but still has top management close enough to the grass roots to know what is happening, and to retain ultimate control. On the other hand, a flat structure could be used to retain strong centralised control.
2. A *tall structure* could be regarded as decentralising, in that it allows for authority to be dispersed. On the other hand, it could be regarded as centralising, as it could restrict degrees of authority at each level.

Span of control

This concept was discussed in the last section, where it was shown that wide spans tend to go with flat structures and narrow spans with tall ones. A few additional points are appropriate here.

The span of control concerns how many subordinates can be effectively managed. A number of factors are influential:

- preference of manager to devolve authority
- ability of subordinates to self-manage
- effectiveness of communications in the organisation
- efficiency of administrative procedures and support staff.

Again there are conflicting conclusions which might be drawn. Well-qualified professional staff should require less supervision, which would imply wider spans of control. But, because of the complexity of the professionals' work a more senior manager might not be able to supervise too many people. On the other hand, unskilled workers may appear to need more supervision, and hence a narrow span of control. But, because of the routine nature of the work, a wider span of control may be possible, especially if some of the 'supervision' is performed by the technology. An example might be in mass production, where machines determine the speed of work, thus reducing supervision, and increasing the span of control.

Delegation

It is argued by Cole (1993) that this is a management rather than an organisational structure issue. However, structure will be affected according to the nature and extent of delegation. Referring to the theory of bureaucracy, delegation is the process by which a part of rational-legal authority is transferred to another.

A distinction is often drawn between authority and responsibility, the first being capable of delegation but not the second. Thus a senior manager may delegate tasks to a subordinate manager, but the senior remains responsible. Therefore, if anything goes wrong, the senior should 'carry the can'. This is clearly not always the case, since senior managers, and politicians, do not always resign their posts, instead of the subordinate, when that subordinate makes a serious mistake.

Delegation can occur in horizontal and/or vertical dimensions. A distinction is sometimes made between line and staff relationships. In theory, a line relationship is vertical in that it is concerned with the operational stages of producing the final goods and services which is the purpose of the organisation. A staff relationship is claimed to be more advisory, and is therefore tangential or horizontal to the main line of the firm's activities. The distinction is of limited value, as roles are often mixed. An individual may have both operational and advisory relationships within the organisation.

The horizontal/vertical distinction may be useful in considering the process of delegation. If the role of a manager or other professional worker is analysed, it may become apparent that there is overload, with the worker not being used in the most effective way commensurate with his or her ability and salary. In deciding how best to respond, the solution may lie in delegating part downwards to more junior 'line' workers and part horizontally to 'staff' administrators. There are many examples, often quoted, where teachers, doctors, managers and others complain that excessive paperwork is preventing them from carrying out their core tasks. This is a case of there being insufficient administrators to delegate to horizontally. This can be as much of a problem as insufficient subordinates to delegate to vertically.

Centralisation/ decentralisation

This issue concerns the dispersion of authority and power throughout the organisation. In particular it concerns who is entitled to take decisions about the deployment of the organisation's resources. There is a strong qualitative aspect to this, since one cannot fully assess the degree of centralisation merely by observing the organisational chart. In principle, a flat structure with a wide span of control could be consistent with any degree of centralisation. It depends on what is permitted at each level, and what rules and procedures are imposed.

Centralisation is favoured by senior managers who wish to retain control over all important decisions. It is also often argued that certain specialist functions are more economically provided centrally. And, of course, central control over finance is often regarded as crucial in ensuring that key objectives of profitability and liquidity are met.

Decentralisation is favoured by some, because it passes more responsibility to the 'sharp end', where costs are incurred, and revenue generated. It should also enable flexibility, so that decisions can be taken more quickly, and in the light of local circumstances and particular customer needs, without reference 'up the line'. This, in turn, frees senior staff, enabling them to concentrate on more strategic matters.

It is argued that most contemporary firms have adopted a multi-divisional structure which is substantially decentralised, but with central control in certain vital respects. This structure also combines certain of the patterns of grouping organisational activities, which were previously discussed.

Multi-divisional organisational structures

These are also known as the M-form of organisational structure. The M-form, or some variant, has become the most common practical expression of organisational structures. It was originally conceived as the only logical way to control large multi-national diversified firms, but the same principles have been applied to much smaller firms in a whole range of industries. In essence, the M-form:

- combines different methods of grouping organisational activities discussed earlier, and
- decentralises a good deal of decision making, while retaining centralised control in the key area of finance.

The purest type of M-form will be described, but there is scope for variation. M-form structures have a top layer of management, such as a board of directors, who

Fig. 7.1 Multi-divisional (M-form) organisational structure

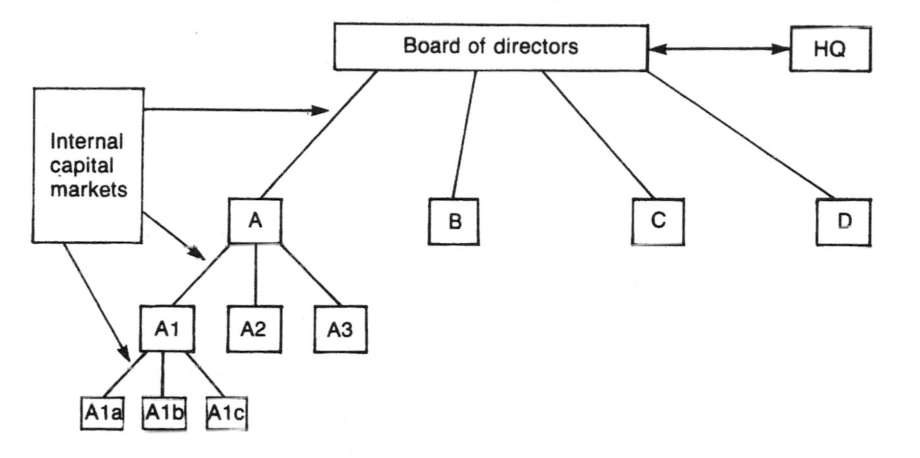

will be serviced by a relatively small head office. Below this will be a hierarchy of divisions, each of which is mainly autonomous. The structure will be expressed in terms of groups of activities. So, with reference to Figure 7.1, assuming the firm depicted is a multi-national manufacturing company, then:

- A, etc. could represent continents
- A1, etc. could represent countries
- A1a, etc. could represent factories each making its own particular product.

Assuming the firm depicted is a national construction company, then

- A, etc. could represent regions
- A1, etc. could represent types of work
- A1a, etc. could represent projects.

There are a number of important features of the M-form of organisational structure:

1. In principle, the numbers of levels in the hierarchy could be quite extensive, depending on how diversified the firm is in terms of products, markets and geographical spread.
2. Each division is responsible to the level above.
3. Each division is set up as a separate company, with its own board and decentralised functions such as production, marketing and personnel.
4. The one function which is never decentralised is financial control.

Although there is a large measure of decentralisation, control from the centre is maintained through financial means. This works through the main board of directors, and head office staff, setting up what is essentially an internal capital market, with the head acting like a bank, and a well-informed one at that. By means of research, it is established at head office what each division should be capable of achieving, in terms of profit. On this basis, capital will be advanced. If profit targets are not met, then capital may be withdrawn, and closure of the division, possibly with transfer of operations elsewhere, is of course the ultimate sanction. It will be recalled that in Chapter 4 it was shown how capital markets deter middle managers and production workers from pursuing objectives other than profit maximisation. The internal capital market does a similar job, but possibly more effectively. This is because it is closer to a perfect market situation, in that the directors and their head office team have better information.

M-form organisational structures are therefore characterised by operational decentralisation, with financial centralisation. The pattern described can be regarded as a pure form. Some firms may resemble this pattern quite closely, particularly holding companies who are more likely to add or subtract from their holdings, in accordance with financial factors. However, it is more common for firms to adapt the model to their own circumstances. For example, some firms will centralise functions such as personnel, or research and development, as well as finance. Furthermore, although each division may be allowed a large measure of tactical freedom, the centre may retain more control over strategy. The application of the M-form structure to construction firms will be further examined in Chapter 20.

Communications in organisations

This is, of course, a major topic in itself and will be dealt with only briefly here. Communications can be formal and/or informal with the latter capable of being very effective indeed. After all, formal committee meetings can often be frustrating, whereas an ad hoc discussion over a beer, without minutes, can result in more decisive action. Many managers, particularly those who favour delegation and decentralisation, are keen to foster informal communications.

As for formal communications, these relate to the formal organisational structure. They are reporting mechanisms designed to ensure that instructions and feedback are transmitted. For effective communications there needs to be:

- the appropriate media of communication, written or oral
- clear lines of, and forums for, communications, be they lines of command or committee structures.

Communications in an organisation can be:

- vertical, such as through oral instructions, reports, presentations or briefings
- horizontal, such as through a working group consisting of peers.

Of course, some means of communication are horizontal and vertical. For example, committee meetings may well have members at different levels of the organisation. Therefore any information communicated will be received by members at the same, and different, levels in the hierarchy.

Finally, the danger of ineffective communication should be realised. Managers should ensure that any barriers to communication are overcome, particularly where these derive from deep and damaging emotions such as fear or lack of trust. Furthermore, communication can be rendered ineffective by having too much of it. Many people can testify to the excess of memos and reports which land on their desks. A good many of these are relatively unimportant, and if the habit of not reading them is adopted, then something which is important will be missed.

Summary

This chapter has considered some of the many facets of organisational structures. Organisations will structure themselves and set up lines of communication which they believe will best help them to achieve their objectives. Starting from a theoretical basis, the first area for discussion was the nature of bureaucracy. The various options which organisations have for grouping their activities were then examined — by function, product, location. There followed a discussion of certain topics related to hierarchies and relationships in organisations — levels of hierarchy and span of control, delegation, centralisation and decentralisation. This led to consideration of the multi-divisional, or M-form, of organisational structure. This has emerged as the most common structure, exhibiting a large measure of decentralisation, but with centralisation through strict financial controls and an internal capital market. Finally, brief consideration was given to aspects of communications in organisations.

Further reading

The literature on organisational structures is, of course, quite extensive. For a standard treatment, see Cole (1993), ch. 6, 24, 25, 28 and Dixon (1991), ch. 4. For a more individual approach, see Handy (1993), ch. 9. Finally, a useful summary of the contributions to management thought by certain writers can be found in Pugh and Hickson (1993), Pt 1.

Part Two

Functions of management

8 Processes of management

Introduction

The overall purpose of management is to help the organisation achieve its objectives. For the firm this means achieving profitability and liquidity, thus guaranteeing survival. Therefore, one way of analysing management functions is in terms of maximising revenue and minimising costs, thus maximising profits. This is to occur within an appropriate time-scale, so that liquidity is maintained. This bears a strong relationship to management practice since:

- generating revenue is an important aspect of marketing management, to be considered in Chapter 9, while,
- controlling costs is an important aspect of production management, to be considered in Chapter 10.

These two functions are supported by tasks such as:

- financial management, to be considered in Chapter 11, and
- personnel management, to be considered in Chapter 12.

Thus, Part Two of this book discusses the main practical aspects of management, with this chapter forming the basis.

Although the headings of 'marketing', 'production', 'finance' and 'personnel' describe tasks within the organisation in a functional sense, they do not describe the processes of management in a more general sense. The usual starting point for studying the processes of management is the well-known work of Fayol, which will be considered in this chapter. Alternatively, a behavioural approach may be adopted which studies what managers actually do in practice, and how they spend their time. This is typified by the work of Mintzberg, and will be more fully discussed in Chapter 13, on management behaviour.

In a book about management in construction, it is appropriate to recognise the differences between management generally and the management of projects. This will be introduced in this chapter, as a basis for more detailed studies in Part Four of this book, 'Management applied to the construction industry'.

Fayol and the processes of management

As mentioned, the usual starting point for considering the processes of management is the work of Henri Fayol. Starting as a mining engineer, Fayol spent the whole of his working life in a single organisation. His book, titled in an English translation as *General and Industrial Management*, was published in 1919, and is regarded as the first attempt to derive general principles of management. This approach contrasted with the contemporary work of the F.W. Taylor school of scientific management, which concentrated on production management.

The starting point for Fayol is to identify *six key activities* of any industrial organisation. From one of these, managerial activities, the *five processes* which make up management are derived. Finally, the *14 principles* of management are developed.

Activities

According to Fayol, any industrial organisation undertakes six key activities, as follows:

1. *Technical* activities, such as production.
2. *Commercial* activities, such as buying resources and selling products.
3. *Financial* activities, such as obtaining long-term and working capital.
4. *Security* activities, such as taking care of the organisation's property.
5. *Accounting* activities, such as compiling and maintaining financial information on the organisation's activities.
6. *Managerial* activities.

It was felt that this sixth group of activities required further analysis, given that a considerable amount was already known about the other five.

Processes

The sixth group of activities above was further analysed by Fayol. It is from here that the five processes which form the basis of Fayol's definition of management are derived. These are:

1. **Forecasting and planning** Looking to the future, and drawing up strategies in respect of the organisation's activities.
2. **Organising** Putting in place the structures which enable activities to take place.
3. **Commanding** Giving instructions, and ensuring they are carried out.
4. **Co-ordinating** Harmonising activities, so that all are working towards common goals.
5. **Controlling** Ensuring that actual performance is in line with planned performance, and taking remedial action, and/or revising plans if necessary.

Principles

From the above five processes, Fayol developed 14 principles of management which are designed to have general applicability to industrial organisations, and possibly to other organisations as well. These are:

1. **Division of work** Specialisation of task which enables expertise to be developed in a narrow band of activity, thus improving efficiency.
2. **Authority** The right to give commands, together with the responsibility arising from it.
3. **Discipline** Managers should command respect, and employees should give it.
4. **Unity of command** An employee should answer directly to one manager only.
5. **Unity of direction** Each group of activities which have the same objective, should have one head, and one plan.
6. **Subordination of the individual interest to the general interest** No single group should be able to deviate from the general interest.
7. **Remuneration** Payments and payment systems should be fair to both employee and firm.
8. **Centralisation** This will be present to some extent, according to such factors as the type of industry and the size of firm.
9. **Scalar chain** A clear line of authority from top to bottom which should in general be followed.
10. **Order** Everything and everybody should be in the most appropriate place.
11. **Equity** Management should be fair in its dealings with employees.
12. **Stability of tenure** Employees should receive proper induction into the organisation, and should not be made to feel that their positions are under threat.
13. **Initiative** This should be encouraged, within the limits of authority and discipline laid down in the organisation.
14. ***Esprit de corps*** Management should build up team spirit and high morale in the organisation.

The relevance of Fayol

Since Fayol, the definition of management based on his five processes has been widely used by other writers, not only as the bases for their own definitions, but also as a structure for numerous management textbooks. Therefore, the core of Fayol's work is still regarded as relevant. However, it is important to consider the context in which Fayol was writing to establish the extent of revision needed in the present day.

Although Fayol attempted to define a universal view it was within a structural bureaucratic model. Thus management is principally based on organisational structures and leans heavily on the Weberian views of bureaucracy, as discussed in Chapter 7. This means that the view is hierarchical, from the top downwards. Many would say that this is inconsistent with modern notions of decentralisation, motivation, and democracy. However, others might argue that some of these issues were only relevant at times of full employment, and more traditional styles of management have reasserted themselves in the 1980s and 1990s, at least in Britain.

An interesting feature of Fayol's work is the high profile given to the human element, for example in principles 7, 11 and 14 — remuneration, equity, *esprit de corps*. This element had little place in the parallel scientific management school in the USA led by Taylor, and on this basis Fayol could be regarded in some respects as a forerunner of the human relations school of thought. This may have something to do with the European economic tradition which differed from the more market-orientated approach of Anglo-Saxon economies. It should also be remembered that Fayol's book was published in 1919, a period of upheaval and revolution in Europe in the aftermath of the First World War. In these circumstances, a more humane paternalistic approach to management may have been considered more appropriate, in order to prevent further revolution, and to safeguard and retain capitalism.

Management processes today

As previously stated, the processes of management as defined by Fayol are still influential today, albeit with some modifications. For example, 'commanding' is no longer regarded as appropriate, in theory if not in practice. Terms such as 'motivating' and 'leadership' are much more to the fore in management literature. Therefore, a typical classification around which management books are often structured is:

- planning
- organising
- motivating
- controlling.

A general description of each will follow, with further consideration in subsequent chapters. In particular, the part that planning and control play in the management of the construction project should be emphasised. This will be considered later in this chapter, as well as in Part Four of this book.

Planning

The importance of planning has already been noted in Chapter 7. Here, it was shown that the predominant organisational structure in modern firms is the multi-divisional or M-form. This largely decentralised form is held together from the centre through financial planning, although many firms adapt the M-form to allow for more wide-ranging central planning, a process known as corporate planning.

Planning involves looking ahead, which implies that forecasting is an inherent part. In the market-orientated firm, making strategic decisions about the desired future market position of the firm is the key to planning. It is useful to consider types of planning, and the planning process itself.

Types of planning
Planning may be classified in terms of:

- level
- time-scale
- degree of detail
- who undertakes it.

At the highest level there is **corporate** planning which is a level associated with the M-form structure. In addition:

- **Strategic** planning is long term, in very broad detail, and carried out at a senior level in the organisation.
- **Management** planning is more medium term, with a greater degree of detail, and carried out by middle management.
- **Operational**, or tactical planning is short term, more detailed, and carried out at supervisory level.

The planning process
The first stage of the process is setting objectives. It will be recalled that objectives of organisations were considered in Chapter 3, when matters such as survival, profitability, liquidity, costs and revenue were discussed. Objectives in specific terms need to be set, appraised and updated when necessary. In setting objectives, account should be taken of the strengths and weaknesses of the organisation, as well as the environment in which it is operating. Thought must be given to how the plans will be put into effect and how they can be changed if necessary. In this respect, the link between planning and control is very important, because control is mainly about ensuring that plans are adhered to.

Organising

This is about implementing plans and, of course, should be taken into consideration at the planning stage. It involves putting into place the necessary structures. Many of the factors concerned with organising were discussed in Chapter 7 on organisational structures, including grouping of tasks, delegation, spans of control, communications.

Motivating

Clearly, if plans are to be implemented, the process of organising, such as putting in place appropriate structures, must be undertaken. In addition, it is necessary to ensure that the people in the organisation perform as required. In Fayol's original scheme of things, the terms 'commanding' and 'co-ordinating' were used.

Although casual observance of some modern organisations may suggest that little has changed, nevertheless greater emphasis in management thinking is placed on seeking the co-operation and commitment of employees. This is the essence of 'motivating', which also embraces matters such as leadership and work groups. These matters will be considered in Part Three of this book, 'Human aspects of management'.

Controlling

Control systems are set up to ensure compliance with plans. The stages of a control system are as follows:

1. Set targets or standards of performance
2. Measure actual performance
3. Compare with target
4. Take remedial action if necessary
5. Revise targets if necessary.

Planning and control are frequently linked together because good planning is the first stage of a control system.

The ability to take remedial action is an important aspect of control, because without this, all that remains is a system of monitoring. This may be useful in its own right, but not to be confused with a genuine control system. For example, in construction projects, a good deal of monitoring of costs takes place at all stages. To qualify as a control system, the information must be in the hands of management in time to take remedial action in the context of the existing project. Where collected information can only be used to help plan future projects, this is solely a monitoring system but can, of course, be very useful. For example, quantity surveyors collect cost data on current projects, specifically to use in the cost planning of future projects.

It should also be noted that when an examination reveals that actual performance has deviated from the plan, it should not automatically be assumed that remedial action is necessary. It could be that the original targets were incorrectly formulated, or that circumstances have since changed, rendering them unrealistic.

With these points in mind, there are certain basic rules about the drawing up of a control system:

1. Targets should be expressed in measurable forms, such as programmes, budgets, and cash flow forecasts.
2. Targets should be realistic and achievable.
3. The system must achieve the appropriate balance of complexity and time, that is, it must yield sufficient meaningful information, but in a time span which enables remedial action to be taken.
4. The setting-up and running costs of the system should be commensurate with the likely benefits.

Concepts of control are used frequently in this book, in a general and in a construction context. For the construction project, planning and control are applied to the elements of the project, including the usual objectives of cost, time and quality, which define a project. For the remainder of this chapter, the processes of management as they apply to projects will be considered, and in particular how these contrast with general management, as hitherto discussed.

Managing projects

The differences between project management and management generally should not be overstated, but there are some clear distinctions which can be made at the outset. For example, general management is ongoing whereas a project

- is unique in character
- is of a temporary nature with a beginning and end
- has a series of deadlines and targets
- requires a project team to be set up, probably of a multi-disciplinary nature, and headed by a project manager.

These characteristics imply certain specific problems of project management:

1. Planning may be more difficult, and realistic targets and milestones are important.
2. Due to the unique nature of a project, it may be difficult to achieve a learning curve.
3. For firms whose business is the delivery of projects, such as building contractors, the size of each project could represent a significant part of their workload, creating a peak loading problem.
4. Because of the uncertainties of many projects, the degree of risk could be very great.
5. Where the project is being undertaken for an external client or customer, the needs of that client and any subsequent user need to be taken fully into account. This problem may be accentuated in the case of an inexperienced client.

A single definition of project management is not easy, but a statement from Lock (1992) gives a good idea of the specific problems, while retaining the importance of the processes of management previously considered:

> The function of project management is to foresee or predict as many of the dangers and problems as possible and to plan, organise and control activities so that the project is completed successfully in spite of the risks.

Types of project

In terms of projects, the main concern of this book is, of course, construction projects. However, project management is emerging as a discipline in its own right, with its own professional body. Much of what has been discussed in the previous section applies to all projects, construction or otherwise. Lock classifies projects under four main headings:

- Civil engineering, construction, petrochemical, mining and quarrying projects
- Manufacturing projects
- Management projects
- Research projects.

Civil engineering, construction, etc. projects

A major feature of these projects is that the production phase is carried out on a site away from the contractor's office, which creates potential problems in terms of weather conditions and transport. As will be discussed in Part Four of the book, building techniques have developed to allow for more prefabrication in factory conditions, while increasingly sophisticated plant is available to facilitate civil engineering projects. Even so, these only help with some of the problems — transporting large components and heavy sophisticated plant may create problems of their own.

These projects are capital investments, often with major sums involved, and are normally carried out by a contractor for an external client. The problems of such contracts are well known in the construction world. With examples such as the Channel Tunnel in mind, many projects are now large enough to make some kind of joint venture or consortium necessary in order to pool resources and share the risks. Because of the risks involved, both technological and contractual, special attention must be paid to project planning and control.

Manufacturing projects

These include manufactured items, usually of an individual nature such as ships, machines and aircraft. Again they often involve major capital sums. They may be built to order for a customer, or they may be internally generated perhaps as a prototype for, say, a new aircraft. Where built for a customer, many of the issues common in construction projects will also apply to manufacturing projects. An example of such a problem may be arrangements for stage payments to the contractor.

One difference from construction projects is that much of the work may be carried out in the premises of the manufacturer, allowing for greater control over the process. However, this must be qualified by the fact that manufacturers increasingly rely on outside suppliers and subcontractors for proportions of the work. In any case it is likely that some of the project will be undertaken away from the manufacturer's premises, for example, the installation and commissioning of new plant in the customer's premises.

Management projects

At some time, most firms will need to undertake a project of a management kind. An example is a feasibility study leading to a restructuring of the organisation. A physical project may be the relocation of the firm's headquarters. Other examples may be all the activities leading to the installation of a new piece of plant or a new information technology system, or the mounting of an exhibition to publicise the firm's products.

In some cases the management of these projects may be carried out in-house, depending on whether such projects occur at reasonably regular intervals. In other cases, project management expertise may be bought in through the use of management consultants or marketing agencies.

Research projects

Pure research is characterised by its uncertain outcome. This means a good deal of money can be spent with no guarantee of a return. Nevertheless, research into new products, techniques, materials and markets is an important part of many firms' activities. Arguably, a strong research base is one of the soundest investments for growth and success. However, many firms are open to criticism for treating research expenditure as a cost, or burden, rather than as an investment.

Because of the uncertainty of outcome, normal project management methods may be harder to apply, but clearly some control must be exercised to prevent costs from escalating, or time from passing, without the prospect of results.

Objectives of projects

The 'cost–time–quality' triumvirate is now well established as a way of defining the objectives of a construction project. Targets in each of these three areas will be set by the client, and the project team will be expected to meet them. In the contexts of projects generally, each requires some further definition.

- **Cost** This may be defined as a budget sum or a cost limit, and the customer will expect it to be adhered to. Failure to do so may result in a reduction in profit and/or a reduction in the return on capital if the project represents the creation of an asset. Even where the profit motive is not crucial, there will still be a need to control costs. In the case of a public project, the emphasis will be on social costs (and benefits). Regular control of costs is necessary. If the costs run too far out of control, then the project may have to be trimmed back or even abandoned, with potential significant losses.

- **Time** Projects have a beginning and an end, and external customers will expect to take delivery by a specified date. This is because the project is often a capital asset such as a building or a machine, which is required for use in the course of normal business. The contractor will require completion in order to hasten final payment, thus improving profitability and liquidity. Therefore, effective time management through project planning and control is very important.
- **Quality** This may be defined either in terms of 'specification' or 'performance', and is undoubtedly the most difficult of the three areas to assess for satisfaction. Not only is it less readily measurable than cost and time, but also quality problems may only become apparent later. Control of quality was traditionally an 'after the event' activity undertaken by inspection and rejection of substandard components. However, the trend now is towards quality management systems to be in place, with the objective that quality of a defined standard is assured from the outset. Quality management will be further considered in Chapter 10.

The three objectives of cost, time and quality are in a trade-off position with each other. For example:

1. Higher quality will normally cost more and might take longer to complete.
2. Faster completion may cost more, depending on whether additional net resources are required.
3. To achieve lowest cost may entail quality being defined at a lower level.

In each case the customer will need to define the relative priorities. For example, a commercial retail client awaiting delivery of new shop premises may regard time as a priority in order that opening before the Christmas rush can be achieved. In such a case there will be less concern about rising costs because this will be more than compensated for by the additional sales revenue generated by being open for business earlier. Many of these trade-offs will be further considered in Part Four, in the context of the construction project.

Frameworks for projects

As already established, projects have a starting point and a finishing point. Therefore, it is convenient to describe the project within a framework which indicates the stages that have to be undertaken to bring the project to completion.

There is no one universally accepted framework. Different frameworks will be used according to the purpose. One of the best-known frameworks for construction projects is the RIBA Plan of Work. Since this has been drawn up mainly for architects, it is orientated towards the design process. The 12 stages are as follows:

- inception
- feasibility
- outline proposals
- scheme design
- detail design
- production information
- bills of quantities
- tender action
- project planning
- operations on site
- completion
- feedback.

An alternative framework, adapted from Lavender (1990), emphasises the stages of economic decision making in carrying out a project. The six stages are as follows:

1. *The idea for the project* — identification of market or social need; consideration of constraints such as availability and price of land, and planning policies.
2. *Decision to proceed* — feasibility studies to ascertain whether the project will meet the objectives of the client; defining project objectives in terms of cost, time and quality.
3. *Design* — producing a workable solution which meets the objectives.
4. *Procurement* — implementing the project by deciding contractual arrangements, selecting contractor, drawing up documentation, and agreeing price and other contract conditions.
5. *Construction* — the production process, including ensuring that planned objectives are being adhered to.
6. *Occupation and use* — maintenance of the building; its management as an asset; eventual replacement decision.

These are but two examples of project frameworks. As stated above, many variations are possible depending on the purpose.

Summary

This chapter has laid the foundations for studying the practical functions of management in Part Two of this book. These functions will be subsequently applied into the construction context in Part Four. The starting point was the work of Fayol, who developed a comprehensive view of management, in contrast to the parallel developments of scientific management which concentrated on the production process. Although Fayol's perspective was somewhat mechanistic and paternalistic compared to many modern management theories, nevertheless adaptations of Fayol's five processes are still widely influential. The processes of management apply generally. Also of interest, in a construction context, is the management of projects. Therefore, consideration was given to the basics of project management, including the identification of different types of project (construction and otherwise), and to project objectives. Frameworks for examining projects were also discussed. For the remaining four chapters of Part Two, concepts such as planning and control will be applied to management functions such as marketing, production, finance and personnel.

Further reading

Most management books contain material on the processes of management and the work of Fayol, for example, see Cole (1993), ch. 2 and 3 and Dixon (1991), ch. 1 and 3. For further information on project management generally, see Lock (1992), especially ch. 1 at this stage. Other references on project management in a construction project will be given later where appropriate, but see CIOB (1992). For an outline of the different stages of economic decision making in a construction project, see Lavender (1990), Pt 1.

9 Marketing management

Introduction

As stated in Chapter 8, marketing management is primarily concerned with ways of generating revenue, so that the overall organisational objective of profitability can be achieved. Precisely how the firm goes about marketing depends on its orientation. Thus, the market-orientated firm will put marketing at the centre of its activities and build everything else around it. By contrast, the production-orientated firm will start with deciding what to produce, before deciding how to market its products.

In this chapter, marketing management will be considered in relation to the firm's existing and potential future markets, market position, and market strength. This requires consideration of market pressures, elements of marketing, organisation of the marketing function, the four elements of the marketing mix (product, price, promotion, distribution) and marketing research.

The firm and markets

The impact of markets on firms has already been discussed, particularly in Chapter 4, 'External influences on organisations'. The firm is affected by its position in markets for resources such as labour, materials, capital and land. However, the main focus of this chapter is the markets in which firms sell their goods and services. Marketing management concerns the ability of firms to generate revenue in the markets in which they operate, or could operate.

As discussed in Chapter 4, a firm's position in a market may be defined by its degree of monopoly, that is, its ability to mark-up price above costs. This is affected by supply factors, such as the number of competitors, and by demand factors, such as elasticity. Generally speaking, supply factors are of a longer term structural nature and change relatively slowly, while factors such as strength of demand are susceptible to more short-term and cyclical changes. The degree of monopoly is important in determining the firm's position in the payments chain, which, as has been seen, affects profitability and liquidity.

Basic elements of marketing management

In considering this function of management there are a number of basic matters to take into account. These include:

- the extent of **market orientation** of the firm
- the distinction between customer **needs and wants**
- the distinction between **features and benefits**
- the effect of **external factors**.

Each will be discussed in turn.

Market orientation

A market-orientated firm is one which puts satisfaction of market demand at the centre of its philosophy and activities, and builds everything else around it. Therefore, the firm will start by ascertaining how to generate revenue through giving the market what it wants, and only then, deciding how best to produce it at minimum cost. By contrast, a production-orientated firm is one which bases its activities around its preferred products and production process, tending to assume that good products will 'sell themselves'.

It is conventional wisdom these days that firms should be market orientated. This fits well with market theory which suggests that the market is the servant of society, and firms must satisfy customer needs if they are to survive. However, this view can be challenged. For example, if a firm has a high degree of monopoly, it might find it more profitable to decide what it can produce most cost-effectively and then seek to manipulate the market to sell its output. This approach is more likely where the firm's production process is organised around expensive items of capital equipment which need to be fully utilised for cost-effectiveness.

Firms in the construction industry may exhibit different types of orientation. The traditional contractor is production orientated, as the product has been fully defined by the client's design team before the contractor becomes involved. What remains for the contractor to do is to give a price, and then organise production. However, many contractors have become more market orientated by diversifying into markets such as housebuilding and property development. Many have also redefined the contracting role by realising that many clients also require organisational and risk management services, rather than simply production management skills. This trend is indicated by procurement approaches such as construction management/management contracting, and design and build. Consultants have also shown greater market orientation by diversifying into other services. Turning to materials and components manufacturers, however, the position is not so clear. Many firms are market orientated, but many enjoy a significant degree of monopoly. In certain cases, such as kilns for manufacturing bricks, the capital equipment is very expensive and needs to be fully utilised for cost-effectiveness.

Needs and wants

Marketing management requires a distinction to be made between needs and wants. Measuring the extent of people's needs has a degree of objectivity. Arguably, needs are limited to essentials like food, clothing and shelter, and one could survive on relatively little. However, it is wants which are the real stuff of marketing. For example:

- it is not 'food' which is being marketed, but particular kinds of high value added food
- it is not 'clothing' being marketed, but a garment as a fashion statement
- it is not 'shelter' being marketed, but a house as a declaration of status and lifestyle.

This raises the question: In targeting wants, is marketing about giving customers what they want, or is it about manipulating them into parting with their money for things they do not really need? To suggest the latter is to question the principle of market economics that consumers have perfect knowledge and always act rationally.

Features and benefits

It is a fundamental principle of marketing that the emphasis should be on the benefits of a product rather than on its features. The features define what the product is, but what encourages the potential customer to actually take the decision to purchase, is the expected benefits. Taking two of the examples referred to above:

1. Buying expensive ready-prepared food such as Chicken Kiev from a supermarket, rather than the basic ingredients from several smaller shops, yields the benefit of speed and convenience.
2. Buying an expensive jacket with a designer label gives the benefit of 'street credibility' among the peer group.

There are numerous examples in the construction context, and it is important always for the firm to know what it is about their product which could benefit the customer. This can be considered from the point of view of:

- consultants
- contractors
- housebuilders.

Consultants

One of the difficulties with marketing the kind of professional services offered by consultants is that these services are invisible, with no actual product to display. Therefore it is even more crucial to be able to identify the benefits to the client. For example, 'project management' may be the service on offer, but this in itself does not define the benefits which would accrue to the client if the consultant were engaged for this purpose. Instead, emphasis should be given to matters such as value for money, or fewer contractual disputes.

Contractors

As mentioned in the section on market orientation, many contractors have redefined their role by offering alternative procurement methods. In marketing these forms of procurement, emphasis should be given

- not to 'design and build', but to single point responsibility, and to greater certainty of price and time
- not to 'construction management', but to earlier start on site, and to faster completion.

Housebuilders

A major difference from the above is that housebuilders are marketing their products to consumers, rather than to corporate or public sector clients. The market for new housing is not a single market, but segmented in a way which reflects the emergence of the 'housing ladder' in recent years. This means that people no longer buy a house expecting to stay there permanently, but instead anticipate 'moving up' the ladder by selling and buying from time to time. Whether this ladder remains the norm in the face of the 1990s slump, and the emergence of negative equity, remains to be seen. Much may depend on whether people continue to see their house as a potential financial investment, or whether they revert to seeing it as a place in which to live, and a possible financial burden.

Assuming the existence of the ladder for now, housebuilders need to understand the sectors of the market in which they are operating, because varying benefits will appeal. For example, in the first-time buyer's market, low cost and convenience have considerable appeal. Many first-time buyers may not envisage a long stay in the property, and so may not be concerned to put their individual stamp on it. Therefore, there would be considerable attraction in being able to move straight into a property with everything already in place, such as carpets, curtains and domestic appliances.

Existing home owners who wish to move may be faced with the problem of not being able to sell their current property. Some housebuilders are successfully dealing with this by offering potential customers the benefit of buying their existing property. The condition is that the customer purchases a new house at a higher price. In the higher price range houses, greater emphasis may be put on images of luxurious lifestyle and prestige. At all levels of the market, benefits such as value for money, fuel savings and environmental friendliness may all be worthy of emphasis.

External factors

In marketing a particular product or service, the benefits being emphasised may be undermined by some external factor which is outside the firm's control. For example, if the swift delivery of a product is the benefit being marketed, then this can be undermined by traffic congestion, caused by an accident or by a general lack of good transport facilities. An accident is hopefully an unfortunate isolated occurrence. However, poor transport means that attempts to offer a fast service will be unsustainable, even though there is a healthy and potentially profitable demand for such a service.

A particular kind of external factor occurs when part of the responsibility for delivering a benefit is given to someone else — a typical situation in construction, where work is often subcontracted out. The main contractor's ability to deliver the promised benefits to the client can be severely damaged by the inability of subcontractors to perform satisfactorily. This emphasises the importance of careful selection of subcontractors and suppliers.

Organisation of the marketing function

Marketing management as a function has grown out of the more traditional function of sales. There is a great deal of variation concerning:

- how marketing fits into the organisational structure
- who is responsible for carrying it out.

Marketing and organisational structure

It was shown in Chapter 7 that the multi-divisional or M-form is the most common organisational structure for modern firms. In the purest version of the M-form, all activities, except financial control, are decentralised to the divisions. Therefore, there would be no marketing department as such at headquarters. However, in practice this is likely to be modified, with some functions retained at the centre. The balance between headquarters and the divisions will vary, with more decentralisation in the case of a conglomerate. For a manufacturing or construction company, it is reasonable to assume that many day-to-day or short-term marketing activities will be left to the divisions, while matters of long-term market strategy are more likely to remain under central control.

Responsibility for marketing

The main question is whether marketing is a particular skill exercised by marketing specialists, or whether it should be the responsibility of everyone in the organisation. In a market-orientated firm, where marketing is central to its philosophy and activities, it is very likely that most or all employees will be expected to think and act in marketing terms. In a production-orientated firm, it is more likely that the marketing department will be separate and possibly seen as a service department to the main business of producing good products which are thought to largely 'sell themselves'.

While it is not easy to generalise, a model adopted by many firms is one where all senior employees have some responsibility for marketing as part of their normal activities. Depending on the size of the firm, a certain number of marketing specialists may be employed. However, there is always the option of employing outside consultants for specific marketing activities such as launching a new product, or devising a new corporate identity.

The marketing mix

This term refers to the group of variables which make up the firm's marketing activities. It can be defined as it exists at any particular moment, but it is likely to change over a period of time — indeed it is regarded as desirable in a healthy firm that elements of the mix will change. In the long term there could be some significant changes to the mix, in line with the firm's long-term market strategy. The four elements of the marketing mix are as follows:

- product
- price
- promotion
- distribution.

Each will be considered separately.

Product

A key marketing decision is the range of products and/or services which should be offered for sale, bearing in mind always that it is benefits, rather than features, which lead to a sale. Each firm must also decide on the appropriate range of product markets that it wants to be in. In an earlier section, the various sectors of the market for new housing were discussed. The individual housebuilder must decide the sectors of the market in which it is appropriate to operate. Similarly for the contractor, the appropriate mix must be decided — contracting, design and build, construction management, housing, aggregates, property development. Not all firms will come to the same conclusion. For example, in the 1980s some contractors resisted the temptation to enter the property development market although it seemed a sure route to increased profitability for many. Manufacturers also have to decide on a project range — for example, whether to produce a full range of volume motor cars, or to concentrate on smaller more specialised segments of the market.

Apart from the product there are a number of other related matters which fall into this category, such as:

- branding
- after-sales service
- product life cycle.

Branding

Branding can be a powerful marketing tool, especially for consumer goods, and can therefore be utilised by housebuilders who aim their products at consumers. Ownership of a brand name can be extremely valuable, and may be included in the balance sheet as an intangible asset. In some cases a firm may seek to acquire another firm solely on the grounds that it owns certain brand names. Sometimes a traditional brand name, associated in the consumer mind with high quality or performance, may be resurrected for use on a new range of products even though the organisation which originally bore the name has long since ceased to operate. Recent examples include the national brewers using the names of regional brewers who no longer exist, having been taken over and closed down; or Rover introducing a range of sports cars carrying the badge 'MG'.

Many retailers have their own brand name. Originally they tended to symbolise value for money alternatives to the established product brands. But in some cases they have come to be associated in the consumer's mind with high quality in their own right, and may no longer be low price.

After-sales service

All purchasers, especially consumers, have certain legal rights in respect of fitness for purpose and quality of products. However, as part of the marketing mix, many firms will offer additional services which may be of benefit to the customer. This could mean changing unwanted items even though there is no legal obligation to do so. It could also mean much more than this, encompassing maintenance and repairs on or off site. Sometimes this will be automatically included as part of the product, and therefore reflected in the price. In other cases, it may be purchased as an option — a useful benefit for those customers who may be unsure of their ability to take care of sophisticated technical equipment such as computers.

Product life cycle

In determining the contribution which a product will make to the firm's profits, it is essential to have some idea of the life cycle of that product. This varies greatly. Some products seem to continue forever as best sellers, with perhaps only an occasional change of packaging. Sometimes even this is unnecessary. Indeed, it may be resisted by customers, as in the case of Coca Cola. Where a product has a more definable life span, certain stages can be identified:

- *Introduction* Profits are low at this stage since sales are few and costs are high, reflecting development.
- *Growth* Profits are rising as sales increase and unit costs fall. Competition is likely to increase.
- *Maturity* Profits are levelling off as sales rise more slowly due to increased competition.
- *Saturation* Profits are falling as sales stagnate in the face of too many suppliers for the size of the market.

- *Decline* Profits are disappearing as sales permanently decline leading to eventual withdrawal of the product from the market.

The actual time-scale for this cycle could greatly vary, as already mentioned.

Price

Price is an important element in marketing since the quantity of units sold multiplied by price determines revenue. This basic assumption of market theory also holds that price and quantity are normally inversely proportional — as price goes up, quantity sold goes down, and vice versa. The effect of this on revenue depends on the precise relationship between the two, a measure which is known as elasticity. This relationship has already been encountered when considering the degree of monopoly, or the ability of firms to mark up their prices. The elasticity of demand which the firm faces is one of the decisive factors. Clearly firms need to know this when setting prices.

In theory, firms keep their prices flexible so that they can respond to competitive pressures. In practice, too, this may often be true, especially in consumer markets. But for many firms there may be difficulties in achieving price flexibility, since price lists may need to be prepared in advance, and they will be expected to be honoured for a given period. Also contracts for the supply of goods and/or services may be agreed in advance. In construction contracts, prices are agreed in advance, and additional payments may only be possible in limited circumstances.

Setting appropriate price levels is always important, but there are particular times when the process is even more significant:

- when expanding into new products and/or markets
- when cost levels are changing
- when competitors change their behaviour.

New products/markets

This may be new products in existing or new markets, or existing products in new markets. Introducing a new product into an existing market should, in theory, be the least difficult to achieve because presumably customer behaviour is known. However, it is important not to alienate customers, and particular care must be exercised if the new product is a radical departure from what has been offered previously. Customers will expect particular price levels to apply.

Introducing new products into new markets presents the difficulty of knowing where to pitch the price so that it is neither too cheap nor too dear. Ideally the firm would need to know the elasticity of demand so that the price can be chosen to complement quantity in order to maximise revenue. However, it is not quite as simple as it seems, because in the post-launch period the firm will seek to build up its market share so that revenue and profits are maximised over the life cycle of the product.

When introducing existing products into new markets, then similar prices would normally have to be charged unless there is the possibility of price discrimination. This might be the case in markets for services, or where there is geographical separation in markets for products. In the case of services, preferential prices could be charged to new customers as long as this does not alienate established ones. For products, it might be possible to separate home and overseas markets in price terms. In the case of location-fixed products such as housing, different price levels can prevail. However, it is important to distinguish between price differences due to different levels of costs (for example, differences in land costs in different locations), and price differences due to price discrimination.

Changing cost levels

When cost levels rise, the firm is faced with a choice:

1. If price is increased to reflect increased costs, then the effect will depend on how competitors respond, and the magnitude of elasticity of demand. If unfavourable, a loss of market share could result.
2. If price is maintained in the face of increased costs, then there will be some reduction of profit, but this may be compensated for by increased market share and hence increased profits in the longer term. Again the behaviour of competitors and elasticity of demand are important.

Of course costs sometimes fall, and the temptation might be to leave prices where they are in order to increase profit margins. This may make sense in the short term, especially if the firm is emerging from a difficult market position caused by a recession. However, this course of action may sacrifice the opportunity to increase market share.

A similar situation occurs when there is a devaluation of the Pound Sterling. This gives the firm the possibility of increasing the price charged to overseas customers in Sterling without it affecting what they, the customers, actually pay in terms of their own currency. This has the benefit to the firm of increasing profit margins. However, the other option is to maintain price in Sterling terms which should keep margins constant. The effect of this is a price cut to the overseas customer in terms of their own currency which might, hopefully, lead to an increase in market share.

Behaviour of competitors

The implication of the last two subsections is that the behaviour of competitors can be critical. If they maintain prices while the firm cuts its prices, then in general the firm will gain market share. However, much depends on the degree of monopoly, which as has been seen depends on the number of firms competing in the market, and the elasticity of demand.

Promotion

The purpose of promotion is to turn potential customers into actual purchasers. It is recognised that to achieve this a number of stages must be passed, which take the customer from being unaware of the products through to purchase. There are six stages in all, as follows:

- unawareness of product
- awareness of product
- interest in and understanding of product
- desire for product, having understood its relevance
- conviction about value of product
- purchase of product.

Clearly, to achieve a sale, a consistent and persistent programme of promotion is necessary, possibly over a long period of time. While these positive moves are unfolding, there is always the danger that negative forces may reverse the process — 'one step forward, two steps back'. Examples of these countervailing forces include:

- the elapse of too much time in mid-process
- a change of personnel so that the potential customer is now dealing with someone else
- new competitors in the market.

There are several methods of promotion, including:

- **Advertising** This occurs through a variety of media, being the aspect of marketing with which the public is most familiar.
- **Personal selling** This is a more targeted form of promotion which depends on an effective salesforce rather than media advertising of a more general nature. This is particularly appropriate for the marketing of professional services through personal client contacts.
- **Sales promotion** This is designed to increase sales through the use of various incentives. Some, such as special offers and free samples, may be aimed at consumers; others, such as demonstrations and sales material, may be aimed at commercial customers.
- **Publicity** This has some similarities with advertising, but may be free in the case of press releases and the like. Public relations is regarded as an important function, and outside consultants may sometimes be used for this purpose.

It is important to reiterate that with all these methods it is the benefits rather than the features of products which must be promoted.

Distribution

Effective distribution is partly a physical problem, in terms of transport. It was previously stated that offering the benefit of a fast delivery service can be hampered by poor transport facilities. This problem can also affect production management, where there is increasing reliance on the fast throughput of resources.

Another aspect involves the channels of distribution. This refers to whether the producer supplies the customer direct, or through some intermediary. The potential links in the chain are:

- producer
- wholesaler
- retailer
- customer.

The traditional practice is for the goods of a manufacturer to reach consumers via wholesalers and retailers. However, at present there is a greater tendency for shorter chains. For example:

1. Some manufacturers now sell direct to the public — a common occurrence for items such as computers. This is more common in the case of sales to industrial customers.
2. Mail order and bulk wholesale warehouses eliminate retailers from the chain.
3. Some major retailers, such as high street chain stores and supermarkets, eliminate the wholesaler.

In the housebuilding sector, selling direct to the customer is quite common, using promotional devices such as advertising and show houses, although sometimes estate agents are used. In contracting, the position is not quite so clear and the channels between client and contractor depend on the procurement method.

Marketing research

It is important that data are collected and analysed. This helps the firm to improve its marketing effort, both in the immediate period, and in the long term. This emphasises the role of marketing research in the planning process for the firm as a whole. Arguably, for the market-orientated firm, marketing research is the key to successful forecasting and planning. The decisions to be taken, based on marketing research, include:

- which markets the firm should be involved in, both now and in the future
- which marketing strategy should be adopted.

In other words, what marketing mix is appropriate, now and for the future. As competition becomes more global, the need for better information on which to take marketing decisions becomes ever greater.

As with any other kind of research, sources of data can be:

- primary data, which is collected directly through questionnaires, surveys, interviews, etc.

- secondary data, which is collected from published sources.

Primary data is likely to be more reliable and appropriate to the firm's individual needs, but will probably be more costly to obtain.

Summary

Marketing management concerns ways of generating revenue, within both a given market position and any potential future market position. It is important to understand market position in terms of competition and the degree of monopoly. It was shown that successful marketing involves concentrating on promoting the benefits rather than the features of products, and that factors external or peripheral to the main product being marketed can have a detrimental effect. Marketing management is often defined around the four elements of the marketing mix — product, price, promotion, distribution — each of which was discussed in detail. Finally, brief reference was made to the short- and long-term importance of marketing research. Market strategies for construction firms will be considered in Chapter 19.

Further reading

For general reading on marketing management, see Cole (1993), ch. 34–9. For marketing management in the construction context, see: Calvert *et al.* (1995), ch. 10; Fisher (1986), particularly section A; Lavender (1990), application L; and the following articles: Chevin (1993), Eldridge (1990), Fisher (1991).

10 Production management

Introduction

In the study of contemporary management, a great deal of interest is taken in the marketing function, as discussed in the previous chapter. While marketing is very important in helping a firm decide what activities it should be engaged in, and how it should proceed in relation to its customers, it should not be forgotten that it is *production* that actually creates the wealth which customers and society need and want. The early writers on management recognised this, and much of the effort of, for example, the scientific management school was directed towards discovering the best way to produce.

This chapter will therefore consider the principles of production management. As a starting point the importance of production to both society and the organisation will be examined, along with the main factors influencing the production process. The performance of the process is measured by productivity, so the factors influencing this much-discussed variable will also be considered.

The history of production since the Industrial Revolution is concerned with finding ways of improving productivity, and the study of this topic has influenced the development of management thinking in general. In the modern era, research has established a classification of types of production which can assist management in their choice of production methods. In the world of contemporary management there are particular topics which receive special attention, and some of these will be considered. These frequently emanate from practices used in Japan, and include particular approaches to control of the production process, and quality management. There are, of course, several control techniques available in production management. In the main, these will be considered in some of the chapters of Part Four of this book, where the principles of production management will be further applied to construction.

Production and society

As stated above, it is production which creates the wealth which society needs. Production can be thought of as a conversion process which receives inputs of resources such as labour, materials, components and capital, and transforms them

Fig. 10.1 Production as a conversion process

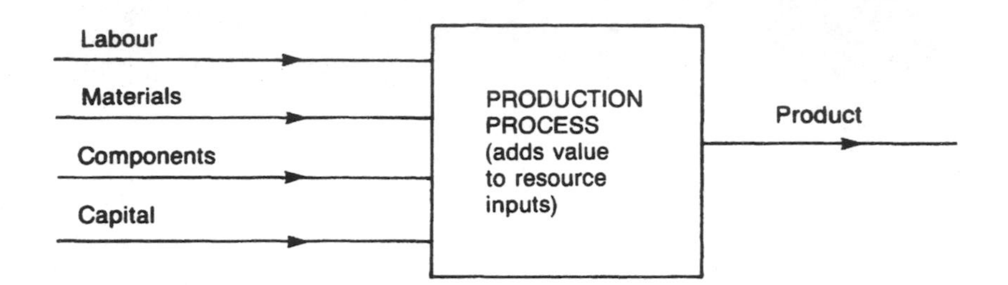

into a product, as shown in Figure 10.1. An important feature is that production **adds value**; that is, the output is more valuable than the sum of the inputs. Thus the production process is the source of economic growth because it has the ability to enhance the resources used.

Although it can be said that production adds value, the quantification of this value is not straightforward, since it could be measured as:

- use value, or
- market value.

Market or exchange value is more easily measurable because it involves comparing the price of the output with the sum of the inputs. Differences between these two measures of value depend on the degree of monopoly in the market. Where this is high, a significant wedge, or mark-up, is driven between the use value based on the cost of inputs, and the market price or exchange value.

Although it is expected that production adds value, and certainly firms normally only undertake production if it adds to their own private value, there are circumstances in which production detracts from the value or wealth available to society. A good example is where production creates pollution as a side-effect. This may add value, or profit from the producer's point of view, but society as a whole has to expend resources clearing up the pollution. This additional cost to society is referred to by economists as *externalities*.

The single production process so far discussed, where the firm buys resources, produces, and then sells its product (a process first shown in Figure 3.1) is not, however, the entire picture. Most completed products are likely to have gone through a number of intermediate stages. Raw materials are first extracted and turned into components, which are then incorporated into products. These products are then distributed and sold. At each stage a firm will be involved, and value will be added, until the final product is received by the purchaser. Therefore, the production of a product requires the involvement of a number of firms.

Production and the firm

It can be seen from the above that the scope of production management is quite wide. It is not merely concerned with what happens on the shop floor, or the construction site, but embraces a whole range of activities. According to Koontz and Weihrich (1990), production management 'has been generally expanded to include such activities as purchasing, warehousing, transportation, and other operations from the procurement of raw materials through various activities until a product is available to the buyer'.

Not only is production spread over a number of stages, but the balance between what is undertaken at each stage changes over time. For example, construction production has changed in that factory made components are now more widely used than traditional materials. Among the consequences of this are:

- more of the value of a building is now created in factory rather than site conditions
- the contractor has less direct control over the production process.

These matters will be discussed more fully in Part Four.

Production has a key effect on the firm's objectives. As established in Chapter 3, to make a profit is a main objective for long-term survival. Figure 3.2 showed that profit derives from revenue and costs. In turn, costs are derived from the 'price paid for resources and the efficiency with which those resources are utilised'. So, in achieving the firm's objectives, production management is concerned with:

- purchasing
- productivity.

Purchasing depends to a large extent on market conditions, while productivity is determined by a range of factors. Before considering productivity in more detail, the general influences on the production process will be identified.

Influences on production

In the social sciences, there have been attempts to derive theories of production. Since production has principally an economic motivation, it might be assumed that economics has a good deal to say about production. In fact it only tells part of the story. The approach taken is to relate output to the inputs through a mathematical relationship called the production function. Thus, output is a function of — that is, dependent on — various independent inputs. An example is:

$$\text{Output} = f(\text{land; labour; capital})$$

Depending on the time period involved, some or all of these inputs can be adjusted to give a different output. This approach accepts the underlying assumptions about competitive markets and homogeneous resources. This could mean that all units of labour are more or less the same in a flexible labour market.

In essence, the above is saying that production depends on the resources used, and is therefore technologically determined. If this were so, production management would be a relatively straightforward matter, with computer programs being used to make choices between different resource mixes. But, of course, this omits the human factor. People are different, therefore labour is variable in qualitative as well as quantitative terms. People are different in terms of skill, education and training, motivation and attitude towards management. Individuals and groups also vary in their behaviour and attitude in different situations. Therefore, it can be said that the outcome of the production process is determined by both

- **technological** factors and
- **social** factors.

As an example, assume that management is thinking of introducing into the workplace a new machine which has the potential for increasing productivity. In technological terms this may not be a problem, but in social terms it could present a host of problems. For example, it may cause disruption to established working relationships, with the consequent effect on goodwill and motivation. The technological/social distinction has several applications at various points in this chapter.

Productivity

As previously stated, productivity is a key element in determining the firm's costs which, in turn, is a key element in determining profit. Therefore, a good deal of production management focuses on productivity.

Productivity is a measure of efficiency in the use of resources. It is a purely physical, non-monetary measure which relates output to inputs, thus:

productivity equals quantity of **output** *divided by*
quantity of **inputs**.

This is conceptually correct but difficult to measure in practice because each of the inputs is likely to be quantified in different units: land in hectares; labour in hours; bricks in thousands; plant in capacity; and so on. The common practice is to take the resource labour as the denominator. Productivity statistics are then expressed in terms of output per unit of labour input. An example is output per person per hour.

This selectivity in the denominator can yield some misleading results unless care is taken to compare like with like. For example, productivity, as measured by output per person, may be increased by the introduction of a new machine. However, this does not necessarily mean that the performance of the workforce itself has improved. In assessing the nature of any improvement in productivity, it is important to look beyond the statistics. Productivity improvements can be due to:

- technological, or
- social factors

and have an effect which is

- short term, or
- long term.

Short-term improvements in productivity

This can be thought of as a 'one-off' improvement which requires no additional resources to be deployed, and no additional investment to be made. A possible scenario would be where a management consultant was engaged by a firm to suggest ways of improving productivity without spending any money! While this may seem unlikely, there may be certain technological and social possibilities.

1. Technological improvements may include better layout of plant; more intensive use of plant; more effective flow of materials.
2. Social improvements may include better systems of supervision; improved motivation of employees through participation schemes such as quality circles.

The point about all these improvements is that they are of a one-off nature and will not give a sustained improvement.

Long-term improvements in productivity

If the same management consultant was given the brief to suggest productivity changes which did allow for additional resources and expenditure, then the agenda now includes investment. It is through a programme of investment that sustained improvements in productivity can be achieved, but, of course, this costs money. Again, there are several possibilities.

1. Technological improvements may come through investment in new capital equipment. A public sector contribution could be through investment in infrastructure.
2. Social improvements could come through a programme of investment in education and training.

If programmes of investment are maintained, there should be a more significant long-term effect, leading to higher productivity and growth.

In interpreting productivity statistics, it is important to understand the reasons for any improvement in order to assess its sustainability. For example, there was an improvement in the figures for the UK in the 1980s. Part of this may have been due to starting from a low base. If the remainder was due to simply making more efficient use of what was already in place, then any improvement may be short lived. If improvements in productivity are to be sustained there will need to be investment — to update capital equipment and to keep the skills of the workforce up-to-date.

The development of industrial production management

Prior to the Industrial Revolution, many people were engaged in primary industries such as agriculture, while others were involved with fledgling manufacturing industries such as textiles. However, these bore little resemblance to manufacturing as it is understood today because production normally took place at or around the home. The Industrial Revolution put in motion a series of changes which have set the pattern for industrial production ever since. The practices on which the Industrial Revolution were founded became the basis of scientific management, the first school of management thought, pioneered by F.W. Taylor and others early in the twentieth century.

Arguably these practices have remained the dominant strategy of management ever since, albeit with some variations and adjustments. To give a brief historical flavour of the strands of management strategy in relation to production, the following headings will be used:

- Scientific management
- Human relations and post-war developments
- Contemporary approaches.

This is a slightly different classification to the one given in Chapter 2 which concerned the management schools of thought in general.

Scientific management

As described in Chapter 2, the scientific management school is often referred to as Taylorism, but to obtain a good understanding, it is necessary to return to the nineteenth century or earlier, when it was normal for families to be largely self-sufficient: growing their own food; making their own clothes and furniture. Surpluses were sold at market to pay for other items. As industries developed, people began to specialise in certain trades from which they earned money to buy the items they needed.

This was the beginning of the **division of labour** which was eventually extended much further. Trades were split into more simplified tasks, famously documented by Adam Smith in the *Wealth of Nations*, who observed that the making of a pin involved 14 operations. Alongside the division of labour, there were two other elements to the industrial production process:

- **mechanisation**, which is the use of machines to assist, or even completely take over, tasks which have been simplified by the division of labour

- **factorisation**, which means that production takes place in a closely supervised setting, that is, a factory.

Although machines and factories tend to be thought of as a package, their influence on productivity can be separated. The potential increase in productivity due to mechanisation is fairly obvious. But factory production, even without mechanisation, is also significant in its own right, because it has the effect of taking people away from the relative autonomy of their home environment, into a closely controlled setting.

The three elements — division of labour, mechanisation and factorisation — together represent the main features of industrial production, and are sometimes referred to collectively as the process of **de-skilling**. Scientific management consolidated this and introduced other elements such as the detailed study of the best way of performing work and monetary incentives. Underlying this is an emphasis on technological rather than social solutions to production management. As stated in Chapter 2, it was argued by Braverman (1974) that scientific management, or Taylorism, has always been the preferred option of management for seeking to maximise productivity through de-skilling. But there have been some other ideas which need to be considered.

Human relations and post-war developments

The human relations school of management thought was described in Chapter 2. The work of Elton Mayo and others highlighted the need for attention to be paid to the social as well as the technological aspects of production, and in particular the role that work groups could play in improving productivity.

This was still a fairly limited approach since it took little account of any individual ability of workers. This was to come in the post-war period with the contribution of sociology and psychology to motivation theory. A large body of management literature emerged in this period, which argued that productivity would be improved if workers were given more scope to reach higher levels of their 'hierarchy of needs'. A solution was to make work more challenging rather than to treat people as merely units of labour. Therefore, good working conditions and more interesting work came to be seen as more effective motivators than simple financial incentives.

These issues will be examined in more detail in Part Three of this book, 'Human aspects of management', but it is worth reiterating a point from Chapter 2. These approaches were prevalent in the period of full employment, and it could be argued that management only adopted them because workers were able to resist scientific management. With less favourable economic conditions, it seems that normal service has been resumed. However, the clock is rarely put back to exactly where it was, and the next section will consider the more recent position.

Contemporary approaches

It was also shown in Chapter 2 that a diversity of schools of management thought have emerged in the post-war period. Among these, contingency theory tends to argue that no universal approach is possible. Strategy will differ between organisations according to the values and beliefs of management, as well as external environmental factors such as the state of the market and economy.

Among the influential ideas to emerge is the concept of **core/periphery** relationships. This means that firms will regard some tasks and workers as core to their operations and others as peripheral. Those deemed to be core will be subject to a different management strategy to those deemed to be peripheral. Core workers might be employed on a permanent basis with good wages and conditions and be 'motivated'. On the other hand, peripheral workers may have insecure employment with poor wages and conditions and will be 'commanded'.

This raises another key concept of recent years, **flexibility**. It is argued, by supporters of the free market, that in changing and competitive global markets, firms must retain their flexibility to respond, so that they can keep costs under control and remain profitable. Flexibility can take many forms, for example:

- workers should be required to undertake a variety of tasks
- different numbers of workers are required at different times
- workers should be expected to vary their hours as required
- it is expected that earnings can vary between periods of time according to factors such as productivity and market circumstances.

The result of this is that traditional employment practices have diminished and there is a much greater preponderance of

- part-time working
- casual work
- temporary contracts.

This has been favoured by the government through their advocacy of flexible labour markets, and by changes to employment law. These matters will be further examined in Chapters 17 and 18.

A particular aspect of flexibility has been the growth of **subcontracting** and self-employment. This, of course, is very familiar in the construction industry, but has become more widely adopted in recent years. Part of this is due to the concentration of monopoly power in many industries so that the large firms can 'farm out' aspects of work to subcontractors and suppliers. Because of their larger numbers they are much more competitive, and the large firm or main contractor can obtain a good bargain — another aspect of the payments chain. Subcontracting is an example of core/periphery relationships as the firm will retain core tasks under their own control, but subcontract out peripheral tasks.

Of course, there is considerable debate about the efficiency and fairness of flexibility. It has been practised in construction for many years. The pattern of development of production methods in construction has followed a slightly different path to manufacturing, and this will be discussed in Part Four of this book.

Types of production

The previous section considered how production management methods have changed over time. This section takes a different angle and examines the technological options for production. This matter was referred to in Chapter 7 where it was stated that the type of technology used influences depth of hierarchy and spans of control.

The research work mentioned in Chapter 7 was undertaken by Joan Woodward. The results of this research identified various forms of technology. The extreme cases are that a product may be individually produced, or produced in a continuous flow, with several intermediate positions. The main factors determining the choice of technology are:

- *physical nature of the product* — complex products may have to be produced individually, while liquids and gases are produced in a flow
- *economies of scale* — some products are cheaper per unit to produce on a large scale
- *size of market* — even where there are potential economies of scale there may be insufficient demand to warrant large-scale production.

Bearing these factors in mind, the following broad options exist:

- jobbing production
- batch production
- mass production
- flow or process production.

Jobbing production

This is production of a one-off nature, either because it is complex, or because it is being produced to a customer's individual requirements, or because it is being built as a prototype with the expectation that eventually it will be produced in quantity.

Production of this kind is characterised by project management as discussed in Chapter 8, and includes the production of individual items of plant and machinery, ships, and, of course, buildings and civil engineering structures. Traditionally this is associated with skilled craft production using basic materials. However, in recent times increasing amounts of components are used which may themselves have been produced in large quantities. The characteristics of this kind of production can be found in construction projects, and will be further considered in Part Four.

Batch production

This is a grey area between jobbing and mass production, ranging from small batches to large batches. It resembles mass production in that it makes use of standardised components, but the numbers produced at any one time are relatively small. Because production can be in small numbers it can be geared towards the needs of individual customers. It is likely that there would be no substantial economies of scale available. Furthermore, the market may not demand large quantities.

Modern housebuilding is a good example of batch production. Typically, small or large estates of houses are built in batches. The elements of each — for example, foundations, brickwork, roofs, doors, and so on — are carried out at more or less the same time for each of the houses before continuing with the next stage of the production process. Although the houses in each batch have similar components they can be varied in accordance with the needs of:

- different groups of customers, some of whom want detached or semi-detached, or different numbers of bedrooms
- individual customers, who will be given a choice in matters such as colour of bathroom suite

Thus batch production gives some of the advantages of standardised production and hence cost savings, with the flexibility of individual customer requirements — quite an advantage for the market-orientated firm.

Mass production

This is a capital intensive form of production which achieves low costs per unit due to economies of scale. Mass production is very much in the scientific management tradition, and is most associated with the name of Henry Ford and the production of motor cars, as referred to in Chapter 2.

Mass production and mass consumption are linked together. This emphasises the point that to take full advantage of the economic benefits there must be a market of sufficient size. Therefore the products of mass production tend to be controlled by multi-national companies who sell in the global marketplace.

Flow or process production

This is similar to mass production in terms of scale and capital intensity. However, whereas mass production is associated with complex units such as vehicles proceeding through various stages of production, and coming off the line, process production is a continuous flow with the output tending to be less complex units, liquids and gases. A good example is the petroleum industry. Also many products of the construction materials industries appear to fit this description — continuous brick kilns and steel section rolling mills being examples. Thus, process production is associated with continuous operation over a lengthy period of time.

Issues in contemporary production management

There have been some changes in production management in recent years. It is sometimes argued that the dominant mass production form of much of the twentieth century, known as Fordism, has been replaced by more flexible market-orientated approaches closer to batch production, and sometimes called post-Fordism. It is not clear whether this is entirely true, but there are other matters which are also influential, some of which are associated with Japanese industry. This has come about partly because of the success of the Japanese economy itself, and partly because many Japanese firms have set up production capacity in the UK, bringing many of their practices with them. Two of the practices which will be considered here are:

- logistics in production
- management of quality.

Logistics

This refers to the whole business of the flow through the production process — that is, the supply chain. This obviously starts with the raw materials, but emphasis is placed on the relationship between the manufacturer and the suppliers of components and materials. The traditional approach is that manufacturers hold large stocks of materials and components ready to incorporate into the production process, and stocks of finished goods awaiting delivery to customers. Not only does this take up space, but it represents a tying up of capital, thus increasing the need for short-term finance. As discussed in Chapter 6, this can represent a considerable cost, and in particular can be damaging to liquidity.

An approach advocated by Japanese firms is **just-in-time** (JIT) systems. This means that suppliers must deliver materials and components exactly when required at fairly limited notice, thus eliminating the need for the firm to carry stocks. For this to work there needs to be a close relationship between manufacturer and suppliers, and a reasonably steady workload so that the latter can effectively plan their own schedules. Without this stability, the problem of holding stocks would simply be passed down the line to the suppliers. This may give cash-flow advantages to the manufacturer, but there is no improvement to the process as a whole, and the system may give rise to a lack of trust. This is similar to the relationship, which sometimes exists in construction, between main contractors and subcontractors or suppliers.

The system being described here means that everything flows through the process as quickly as possible, so that nothing is idle for long, thereby attracting the need for short-term finance. The principle can be extended beyond manufacturing into, say, retailing where the large supermarket chains have become expert at ensuring a fast flow of goods due to effective purchasing, warehousing and delivery to stores.

The system sounds fine in theory, but there are some reservations:

1. As mentioned above, the system must be operated in good faith, otherwise it simply becomes a vehicle for large firms to take advantage of their degree of monopoly to put pressure on smaller firms, particularly on their cash flow.

2. Fast flows of goods require an efficient transportation system, which is a matter of some concern, certainly in the UK.

Quality management

Quality control is a well-established aspect of production management, but quality management is a more broad-based approach to quality, which embraces the entire way in which the firm operates. Some comparisons between the two approaches will be made shortly, but first some basics. Various terms are in use which overlap, examples being 'quality assurance' and 'total quality management'. Many in the construction industry will be familiar with British Standard 5750. Firms who have achieved quality assurance certification under this standard have shown that they have developed the necessary framework and procedures to enable a quality management system to be established.

The Japanese total quality management approach is meant to permeate the entire organisation. It is consistent with other previously discussed concepts such as market orientation, just-in-time systems, and establishing close and trusting relationships with suppliers and subcontractors. In fact a total quality management system implies that everyone treats whomever they deal with, inside and outside the organisation, as a valued customer or supplier, and expects the same in return.

To understand what this means in practice, it is worth making some comparisons between the total quality management (TQM) approach and traditional quality control (QC).

1. TQM is a 'before the event' approach which aims to design out problems with the product and production system; QC is an 'after the event' approach which checks to see whether the finished product has conformed.
2. TQM expects to get everything right first time; QC expects that there will be some faults which need to be corrected.
3. TQM is regarded as an investment in a system designed to prevent problems, increase productivity, and hence reduce costs and increase profits in the long term; a QC system is regarded as a cost of production which has to be borne and is likely to be reflected in increased prices.

Japanese approaches to production management have been influential to date, but their successful implementation does require a non-confrontational approach, inside and outside the organisation. Whether UK industry, including the construction industry, can adopt this approach remains to be seen.

Control of production

There are a good many techniques available to plan and control production. These cover many aspects of cost, time and quality and include programming through bar charts and networks, and productivity techniques such as work study. Some aspects of financial control are dealt with in Chapter 11, but in the main these techniques will be considered in Part Four of this book, in relation to the construction process.

Summary

Production concerns adding value to resources to create the wealth which society needs. Production management for the firm is about achieving this in a cost-effective and profitable way. There are several influences on the production process, which can be classified as technological and social. The performance of the production process is measured by the efficiency of converting inputs to outputs, normally called productivity. This is difficult to measure in practice, so is normally related to units of labour input. In assessing the productivity performance of a given production process, or of an industry, or of an economy as a whole, it is important to distinguish between improvements of a one-off nature and those of a sustainable nature. The former might be achievable within present resources, while the latter is likely to require investment in additional resources. The industrial process has developed since the Industrial Revolution. This introduced the division of labour, mechanisation and factory production in a closely supervised setting. This has been continued in the twentieth century with scientific management and mass production arguably remaining the dominant forms, although amended at times by human relations and motivational approaches. It has been argued that in recent years more flexible market-orientated approaches, based on batch production, have also been used. Partly this has been due to the influence of Japanese industrial methods which have encouraged producers to reconsider their approach to matters such as the logistical flow through the production process and the management of quality. Some of the techniques for planning and controlling the production process will be considered later, particularly when the construction process is examined in Part Four of this book.

Further reading

For an examination of the economic principles underlying production management, see Lavender (1990), chs 1, 2, 4 and application M. For a general management approach to production and productivity, see Koontz and Weihrich (1990), ch. 20, and Cole (1993), chs 30–2, 41–3. For a radical interpretation of production, especially scientific management and Taylorism, see Braverman (1974). The book by Braverman opened a major debate about the nature of work and management strategies in the production process. The ensuing literature includes Friedman (1977), Edwards (1979) and Wood (1982).

Other reading relating production management to construction will be given in Part Four.

11 Financial management

Introduction

This chapter can be thought of as being in two sections. The first follows directly from Chapter 6. In that chapter, on financial structures, various financial statements were examined, and it was indicated that certain information can be drawn from them which is useful to management and to other interested parties such as investors and customers. This information includes financial ratios, such as profitability and liquidity, which aid assessment of the firm's financial performance. Financial data is usually more useful if comparisons across industries are made.

The second part of the chapter explores areas of financial decision making which are based on information which is only available within the organisation, rather than information which is published externally. This section starts by considering financial planning and control, or budgets, before examining aspects of short- and long-term financial decision making such as costing and capital budgets.

Interpretation of accounts

Information taken from published accounts is, of course, available to anyone. In this respect, there are four main types of ratio which are commonly referred to:

- profitability
- liquidity
- solvency
- efficiency.

Examples of each will be briefly considered.

Profitability ratios

These are the ratios most commonly used as the yardstick of organisational performance. A figure for profit alone would not be very informative, as the same profit made by firms of different sizes cannot be regarded as equal performance. Therefore, profit must be related to the amount of capital employed in the firm, the latter being taken as indicative of size.

Examples of profitability ratios are:

- **return on capital employed** (ROCE) — that is, profit divided by capital employed; this measures overall performance
- **profit margin** — that is, profit divided by sales; this is a measure of market power.

The ratio ROCE is at the top of a hierarchy of ratios, and will be considered separately.

Liquidity ratios

These ratios can be extracted from the published accounts and are used to assess whether the firm is likely to run into cash flow difficulties. Again the two main examples are:

- **current ratio** — that is, current assets : current liabilities, and is recommended to be 2 : 1.
- **acid test** — that is, liquid assets : current liabilities, and is recommended to be 1 : 1.

A number of points need to be made:

1. The recommended ratios cannot be applied across all sectors. For example, firms operating in the retail sector can rely on a more steady cash flow than other sectors, and can therefore safely survive on smaller ratios.
2. Having a high ratio may make the firm ultra-safe in liquidity terms, but it also means that insufficient use is being made of its financial resources.
3. The definition of liquid assets can vary. It should be that part of current assets which can be *easily* turned into cash in an emergency. Ideally this should be restricted to cash-in-hand and in bank.

Solvency ratios

Solvency has similarities to liquidity but there is an important difference. Liquidity ratios give warning of the possibility of firms encountering difficulties due to insufficient short-term or working capital. Solvency ratios give warning of the possibility of firms encountering difficulties due to a problem with capital structure. In this case the problem could be high gearing, that is, a high ratio of fixed interest capital such as loans, to variable interest capital such as ordinary shares or equity. Capital structure and gearing were more fully discussed in Chapter 6.

In the present context, the dangers of high gearing will be discussed. In essence, changes in certain variables can exaggerate or 'gear up' a problem, due to the high burden of debt repayment that must be made irrespective of circumstances. These variables include:

- increase in **interest rates**, leading to greatly reduced funds available for investment and other purposes
- decrease in **profit**, which might dramatically reduce the amount available for shareholders, leading to reduced dividends, selling of shares and possible takeover.

When one or more of these variables becomes unfavourable, great pressure can be put on the survival of firms. For example, rising interest rates and the collapse of the property market in the early 1990s was very damaging to many firms in the construction and property industries.

Efficiency ratios

There are certain ratios which give an indication of whether a firm is managing its financial affairs effectively. If the ratios are unfavourable, it may be due to a problem of internal administration which could be corrected. Alternatively it might be due to an external problem, such as customers not paying their bills because they are facing financial difficulties. It is an important general rule of financial management that if events are not as expected, then the reason must be investigated so that corrective action can be taken.

Examples of efficiency ratios include:

1. Debtors' ratio, which is debtors : average daily credit sales, and is a measure of the average time taken to collect debt.
2. Creditors' ratio, which is creditors : average daily credit purchases, and is a measure of the average time taken to pay debt.
3. Stock turnover, which is cost of goods sold : average stocks of finished goods, and is a measure of the extent to which working capital is tied up in stock.

A note on the use of ratios

Ratios can provide useful information, but must be used with caution. On their own they may have limitations. For example, it was mentioned above how important it is to ascertain why a ratio is of a certain value. It is widely accepted that ratios are of most benefit when used for comparison purposes with, for example:

- previous performance of the firm, say, over a number of years
- other firms in the same industrial sector
- expected or target performance.

With this in mind, financial ratios can be of value.

Return on capital employed (ROCE)

This is the ratio which is often regarded as the best indicator of financial

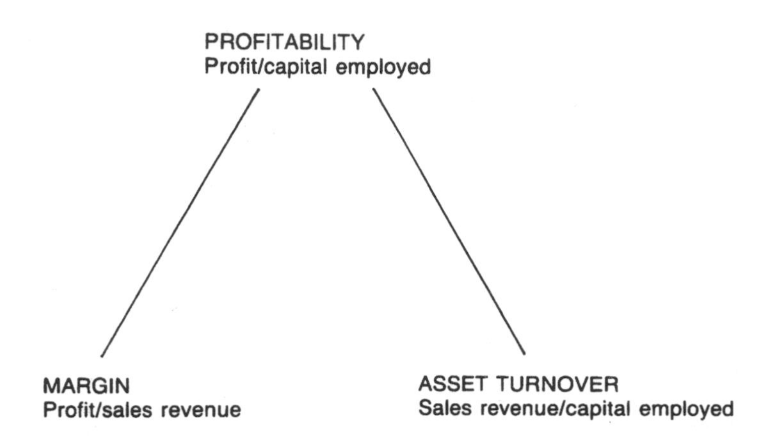

Fig. 11.1 The ROCE pyramid

performance. It relates profit to a measure of the firm's size. Accountants may argue about the precise meaning of 'profit' and 'capital employed', but the purpose here is to consider the practical implications of this ratio for the financial management of firms in different industries.

ROCE heads a pyramid of ratios. This is a similar concept to the 'profit pyramid' described in Chapter 3 (Figure 3.2). It was shown that although profit is the final objective at the head of the pyramid, it is the second and subsequent levels of revenue and costs which help diagnosis of the true position. Similarly, with the ROCE pyramid, illustrated in Figure 11.1, it is the two subsidiary ratios which are informative.

To understand the importance of this, imagine two firms:

1. Firm A is in a highly competitive industry which, by definition (low degree of monopoly — see Chapter 4), means that profit margins are low.
2. Firm B is in a more monopolistic industry where profit margins are higher.

Does this mean that firm A is less profitable than firm B? The answer is *not necessarily*, because profitability depends not only on *profit margin*, but also on *rate of asset turnover*. So, if firm A puts less capital into the business than firm B, but makes it work much harder, then it can be equally profitable.

Therefore, ROCE — that is, profit — divided by capital employed, is derived from:

- **Margin**, which is profit, divided by sales revenue (that is, mark-up) expressed as a percentage
- **Asset turnover**, which is sales revenue, divided by capital employed.

There seems to be a trade-off, because:

- in competitive markets, margins tend to be low, so a high asset turnover is needed to compensate, and to maintain profitability
- in more monopolistic markets, margins can be higher, so a high asset

turnover is less necessary for profitability, and in some cases may be difficult to achieve.

To aid understanding of the ROCE pyramid, examples will be taken from manufacturing, retail and contracting industries. There is a problem of defining what is meant by 'capital employed'. Accountants may seek an objective definition, suitable for all firms, so that a consistent figure for 'profit' can be derived. However, for the purposes of understanding financial management, it is more appropriate to consider those significant assets which form the basis on which the firm's product and profit is generated. These will vary according to industry.

Manufacturing

The basis on which a manufacturing firm generates profit is its fixed assets such as plant, machinery, factory buildings. It therefore seems appropriate to calculate a measure of profitability based on those fixed assets. Fixed assets are long lived, so by definition annual asset turnover will be small, as it takes some time to work those assets to their full extent.

Therefore, for a manufacturing firm to be profitable, it must achieve much higher margins. This should be possible because, in many spheres of manufacturing, large investment is required which restricts the number of firms, thus encouraging higher degree of monopoly and margins. Even so, it is not advisable for the firm to simply rely on monopoly power, but instead it should strive to improve margins. Manufacturers have often been criticised for being too production rather than market orientated, thus missing the opportunity to improve margins and profitability.

Retail

In retailing, the basis of profitability is not the working of fixed assets, but the buying and selling of stock. Therefore, it may be logical to treat stock as the main asset, or capital employed. This does not mean that the retailer has no fixed assets, but that they often play a subsidiary role. More often than not:

- the goods to be sold are made by others, even the supermarket's own brands
- transportation of goods may be subcontracted out
- the buildings where goods are sold may be little more than basic shells.

Therefore, profits are generated in accordance with mark-up on stock (margin), and how quickly stock is sold and paid for (asset turnover).

There are various kinds of retailing, for example:

1. Low-value items such as basic foodstuffs and inexpensive clothing. These are relatively easy markets to enter, so competition tends to be high and margins low. Therefore, to be profitable a high stock turnover is necessary. This is not impossible, and many entrepreneurs have become very

successful this way.

2. High-value items such as washing machines, cars, fashion clothing, etc. Stock tends to turn over much more slowly so higher margins are necessary. The higher margins allow for greater discounts in times of stock clearance or seasonal 'sales'.

Some retailers have attempted to capture the best of both worlds, the large supermarket chains being a case in point. Traditionally they concentrated on a fast turnover of low-value items — pile it high and sell it cheap! But during the 1980s there was an attempt to move up-market by selling higher value-added items just as quickly. This refers to 'enhanced' foods which could be sold for an additional margin over the basic food, presumably with the attraction of greater customer convenience.

Contracting

Margins in contracting have traditionally tended to be low, due to ease of entry into the market, and high competition. Therefore, high asset turnover is required for profitability. But what are the contractor's assets on which profits can be generated? There is usually little in the way of fixed assets. Plant is often hired, there is limited stock, and overheads may amount to a small office, and few staff.

This means that a little capital is expected to go a long way, with a relatively large number of contracts being financed from this limited capital. Therefore, there is high asset turnover, with the 'assets' effectively amounting to the working capital tied up in the contracts. This is really money which is temporarily in the firm's cash flow. For asset turnover and survival, let alone profitability to be maintained, it is essential that the payments chain is not broken in any way that is detrimental to the contractor. This is a precarious situation which partly explains why the failure rate among contractors is high. It also helps explain why many changes have taken place in the construction industry, such as the emergence of large diversified firms, who do not rely solely on the cash-flow cycle in contracting for generating profitability. Many of these issues will reappear later in the book, especially in Part Four.

Financial planning and control

This second part of the chapter considers how firms manage their finances. Again it is the principles which are of importance. Details such as numerical examples can be found elsewhere (see 'Further reading').

Planning and control are two of the management processes described in Chapter 8, and involve:

- setting targets
- monitoring progress

- taking remedial action, if necessary
- revising targets, if necessary.

The main instruments of financial planning and control are **budgets**. Planning involves the preparation of budgets for different purposes within the total budgetary process. These are then subject to budgetary control.

The total budgetary process is, of course, carried out every year and is constantly updated, but it must fit within the overall objectives of the organisation, both short term and long term. If, for example, there is a long-term strategy to break into a particular market, then this must be planned for now, by making an allowance in this year's budget for the necessary expenditure on items such as equipment and staff training.

The question for the firm is: Where should we start the budgetary process? The answer depends to a large extent on whether the firm is:

- production orientated, in which case it will start with the production budget and then decide how to sell its products, or
- market orientated, in which case it will start with the sales budget and then decide how to organise production in the most effective way.

It is generally argued that the firm should now be market orientated. This means that the first priority is to decide what the market wants in both the short and long term, and then build everything else around these decisions.

As mentioned, there are many kinds of budget. By combining information from different budgets it is possible to derive forecasts of financial statements such as the balance sheet and profit and loss account. These are useful when seeking funds from investors and banks. Budgets may be for any convenient time period, but a monthly breakdown for a period of a year is typical. Some of the more significant budgets will be briefly described.

Sales budget

This is essentially a revenue budget, as it will provide forecasts of likely sales of each product or service sold, in terms of selling price and quantity. As previously stated, this assessment of the market is a starting point for the budgetary process.

Production budget

The purpose of this budget is to plan the most cost-effective production programme, based on the expected programme of sales determined by the sales budget. It is important to ensure that production stays ahead of sales, but without building up excessive stocks of either resources inputs or finished goods. Excessive stocks represents working capital tied up, which has implications for asset turnover and hence ROCE.

Cash budget

This budget is for the planning and control of liquidity, rather than the profit forecast (the latter can be derived from the above-mentioned budgets — the sales budget indicates when revenue will be generated, and the production budget indicates when costs will be incurred). However, neither sales nor production budgets indicate when cash will actually flow in or out of the business. It is important to know this, so that overdraft facilities or other short-term finance required may be arranged well in advance. To draw up a cash budget, additional information on payments and credit periods must be known.

Capital expenditure budget

This plans how much of the firm's financial resources will be allocated to investment expenditure. This may be to replace worn out capital, or for new investment, aimed perhaps at strengthening the firm's position in existing or new markets. Decisions taken in respect of this budget will have a long-term effect on the firm.

In summary

Budgets are valuable tools in pursuing the objectives of profitability and liquidity. Another important task for financial management is the control of the cost elements of these budgets. This will be considered next.

Analysis and control of costs

There are many ways of looking at costs. It has already been seen that costs depend on the price of resources and the effectiveness of those resources in use. It has also been seen, in the context of the profit and loss account, that in arriving at a figure for profit, first the direct costs, related to goods sold, and then the indirect costs of the firm, are deducted from revenue. It was also shown, in Chapter 3, that resources used by the firm can broadly be classified as variable or fixed.

It is this latter distinction which forms a common and useful basis for the analysis of costs. Thus:

- **variable costs** depend on the level of production, whereas
- **fixed costs** are incurred whatever the level of production.

Variable costs are usually thought of as attributable to resources such as labour and materials, whereas fixed costs are overheads such as head office expenses which will be incurred irrespective of the level of activity. Therefore, a distinction is often made that variable costs are *avoidable in the short term* if there is no production, whereas fixed costs cannot be avoided. As will be shown later, this distinction can be utilised by management when making pricing and output decisions in the short term.

Although the fixed/variable cost distinction is useful, there are certain provisos:

1. In reality, many costs are neither totally fixed nor totally variable, but lie on

a continuum between the two. Supervision is a good example, because as long as a production capability, such as a machine, is operating, then supervision costs will be incurred whatever the actual level of production. If the machine is closed down, then supervision would be saved, but not the fixed costs such as rent.

2. As the time period under consideration increases, so more costs become variable. Thus head office staff may be a fixed cost in the short term, but if levels of production remain low over a period of time, then management are likely to examine 'fixed' costs or overheads, to see if they can be made 'variable' and hence reduced.
3. Although production labour is conventionally thought of as a variable cost, the individual firm may be loathe to lay off skilled labour, even if the firm is working below capacity. In such a case, the workforce may become, for all intents and purposes, a fixed cost.

The terms 'costing' or 'cost control' can encompass a wide range of items, and brief consideration will be given to:

- absorbtion costing
- marginal costing
- standard costing

each of which has a different purpose.

Absorbtion costing

This approach entails ascertaining the full cost of the goods and services which the firm produces. It may be relatively easy to calculate the variable costs, but the fixed costs or overheads are another matter. This is made particularly difficult if the firm produces a range of goods and services.

One of the main reasons for doing this is to calculate the cost basis to which a profit mark-up should be added to derive price. Therefore, the full cost pricing method describes the textbook approach to construction estimating, where unit rates are built up from labour, materials, plant and overheads.

In more general terms, the cost of a good or service will be the sum of:

- variable costs per unit
- proportion of fixed costs per unit, for overheads which can be specifically identified with the good or service
- proportion of fixed costs per unit, for overheads which cannot be attributable to any particular good or service.

The major problem, of course, is how to allocate the cost of overheads. For the second item, an assumption should be made about level of output of the product to be costed. The third item is more difficult, as assumptions need to be made about the output of all goods and services which the firm produces.

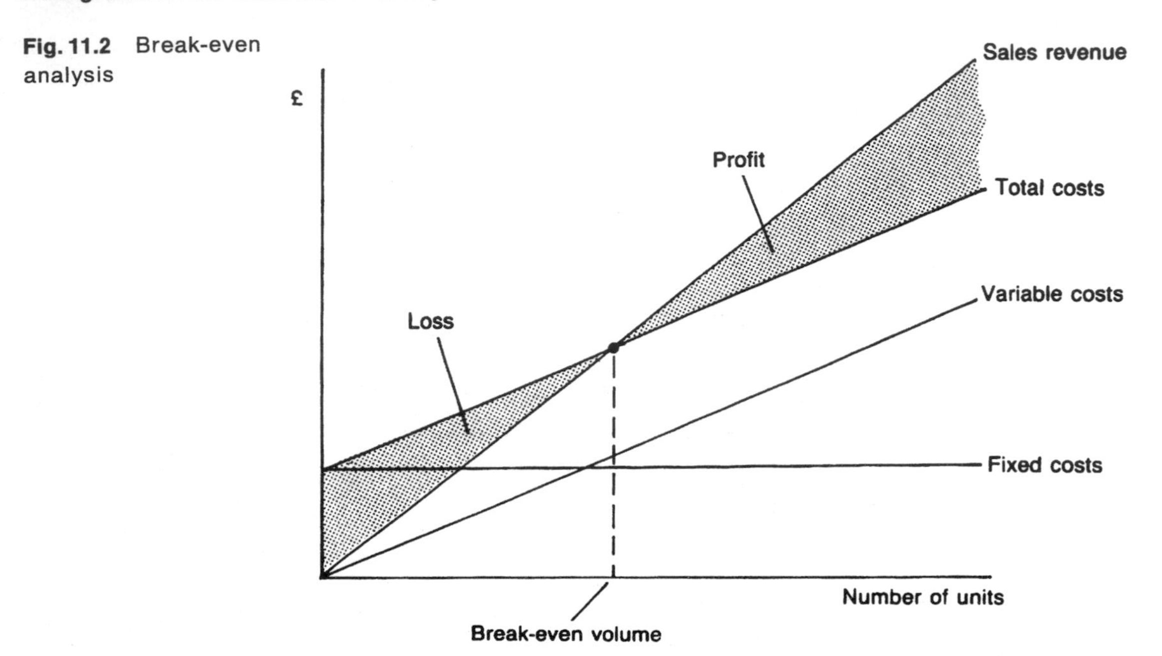

Fig. 11.2 Break-even analysis

Marginal costing

This approach emphasises the difference between fixed and variable costs, thus concentrating the mind of management on those costs (variable) which can be influenced in the short term. Marginal costs are by definition the costs of producing an additional unit. This is equivalent to variable costs. Making the distinction between fixed and variable costs in this way enables management to utilise the following techniques:

- break-even analysis
- marginal cost pricing,

both of which have short-term application.

Break-even analysis

Also known as profit volume analysis, this helps determine the point at which the level of production crosses the boundary between loss and profit. It is a matter of plotting revenue and costs on the same pair of axes to see where the break-even point occurs. Of course information on selling price and costs must be known, so that the revenue and cost curves can be drawn accurately. The application of this technique is shown in Figure 11.2.

Marginal cost pricing

Instead of charging a price which covers full cost plus mark-up, price is based on variable, that is, marginal cost, plus whatever **contribution** can be made in accordance with the state of the market. This is a short-term possibility only, because in the long term the firm has to recover all costs within its pricing, otherwise the capital base of the firm will be eroded and its survival threatened.

Standard costing

This conforms to the planning and control pattern of setting targets, monitoring progress, taking remedial action and revising targets. It is a method of cost control particularly suited to manufacturing, where production is in units. Standard costing also operates within the framework that costs are derived from price of resources and use of resources.

The method sets targets by calculating predetermined or 'standard' levels of sales and cost categories. Progress is monitored by comparing actual performance with the expected standard. Any discrepancy is called a **variance**. If there is a variance then the reasons must be investigated. This can be illustrated by taking two examples.

1. *Sales revenue* If the revenue generated differs from that expected, then the cause may be due to variances in:

   — selling price, and/or
   — quantity sold.

2. *Labour costs* If these differ from the expected, then the cause may be due to variances in:

   — price paid for labour, and/or
   — productivity of labour.

Similar calculations can be carried out for other cost categories.

A general rule about standard costing is that variances can have a price and/or a volume cause. Standard costing is not easily applicable to contracting, since it is not based on units of production. However, the concept of variances is important and widely utilised in cost control in construction.

Long-term financial decisions

The main consideration here is the control of capital expenditure. As mentioned earlier in this chapter, firms are likely to have a capital budget each year, both to replace existing capital and in accordance with their long-term strategy.

Investment, by definition, involves giving up the benefit of the use of resources now in exchange for some greater benefit in the future. Thus, instead of investing in current assets such as stock or debtors, firms invest in long-term capital. The benefits which a firm can expect to receive from an investment vary according to industry, but in all cases they include one or both of:

- **income**, which could derive from increased productivity from a machine, or rents from a property asset
- **capital gain**, which could derive from the enhanced value of the firm, or from the subsequent sale of an asset at a profit.

Fig. 11.3 Cash flow pattern of investment appraisal

YEAR	0	1	2	...	n
COSTS	Capital cost	c_1	c_2	...	c_n
BENEFITS	m	b_1	b_2	...	b_n

$c_1, c_2, \ldots, c_n$ – periodic costs such as maintenance
$b_1, b_2, \ldots, b_n$ – stream of returns such as rents, or contributions to productivity

Capital investment decisions are of two types:

- to assess an individual investment opportunity, to ascertain whether it meets some prescribed criteria such as an acceptable rate of return
- to choose the best option among a range of alternatives.

The **rate of return** will always be important, because this represents the additional benefit required as a reward for giving up current benefit. However, the rate of return is not the same for all investments because it must be set against **risk**. Generally speaking, the greater the risk, the higher the expected return. In many cases long-term funds will only be available if risk is low. The other factor which can affect the return is the extent of **liquidity**, which is the ability to turn an investment back into cash if required. This will depend on the saleability of an asset, and might be difficult for assets such as fixed machinery, but possibly easier for vehicles or property.

Investment appraisal

There are several methods of carrying out an investment appraisal, but all involve *comparing benefits with costs* to ascertain the rate of return. Each investment opportunity will have a cash flow pattern of the type shown in Figure 11.3. The costs will include the capital cost at the beginning of the period (year 0) and any other costs, such as maintenance, in subsequent years. The benefits will normally be a stream of cash flows such as rents or contributions to productivity. The methods of appraisal themselves may be divided into those which do not take account of when costs and benefits occur, such as **payback** and **average rate of return**; and those which do, such as **discounted cash flow**.

Payback
This is straightforward to understand and apply because it simply asks: How long will it take to get the original capital outlay back? This method ignores the eventual total return. So any investment which does not meet a payback period of, say, five years may be immediately rejected. Similarly, an investment which pays back in four years may be preferred to one which pays back in six, even if the total return on the latter is much greater.

Average rate of return
This method takes account of the costs and benefits over the whole life of the investment. The total return is calculated, and divided by the number of years to find the average. However, it does not take account of the timing of costs and benefits.

Discounted cash flow (DCF)
This method takes account of all the costs and benefits, and their timing. The method is based on the time value of money. This means that costs and benefits now are worth more than similar amounts in the future. Of course, this may be due to uncertainties such as inflation, but it is also true because of the rate of interest — money placed in a safe financial investment now will grow, due to accrued interest. This is known as **compounding**. Another way of putting this is that a lesser sum needs to be invested now to accumulate to a given amount in the future, due to compound interest. Therefore, a sum of money in the future is worth less than a sum of money now because it will not have the benefit of accrued interest. Therefore, it needs to be **discounted** back to the present day. Discounting is the inverse of compounding.

When all costs and benefits are discounted back to the present day then a comparison can be made with the capital outlay to ascertain whether the investment yields an acceptable rate of return. If there is a choice of investments, then it can be ascertained which gives the best return.

Which investment appraisal method?

On the face of it, DCF is the only realistic method because it takes account of all the costs and benefits, *and* their timing. However, there are several reasons why something apparently as crude as the payback method may still be valuable.

1. If there are a wide range of capital investment opportunities open to the firm, then payback is a relatively easy way of drawing up a short list.
2. Many investments may be financed by an external source such as a bank, which may insist on a maximum payback period as a condition of the loan.
3. In a period of inflation or other uncertainty, there is a case for saying that sophisticated appraisal methods which attempt to look too far into the future are of limited value. Therefore, it may be preferable to use a simpler method such as payback.

Summary

In this chapter the financial management of firms has been considered. Starting from a review of financial statements as discussed in Chapter 6, information such as financial ratios were examined. The key ratio of return on capital employed was given special consideration in the context of cross-industry comparisons. Following this, the management of the firm's finances were considered, including the budgeting process, short-term decisions such as costing, and long-term decisions such as capital investment appraisal. This general approach to financial management will be placed in the context of construction in Chapter 22.

Further reading

As stated at the end of Chapter 6, there are numerous books on accountancy, including those for the non-specialist. For example, Davies (1992), chs 4, 5, 7, 8, 9, 10, 11, 13 and 19. In addition, for the section on long-term financial decisions, see Lavender (1990), applications F and H.

12 Personnel management

Introduction

This chapter, the final one in Part Two of the book, provides the link to Part Three, which considers various human aspects of management. As stated in Chapter 8, personnel is one of the important management functions supporting the main profit-making functions of marketing and production. It is therefore appropriate to take an overview of the personnel function in Part Two.

The personnel function is not treated with equal importance in every organisation. Even in the same organisation, its importance may vary at different times. The reasons for these differences will be considered. Other differences to be examined are how the personnel function is undertaken, by whom, and how personnel interacts with other functions.

With this groundwork established, some of the normal tasks of personnel will be considered separately, including recruitment and selection, training, job evaluation and appraisal. Some other personnel functions will be briefly mentioned here and given further consideration in Part Three of the book. Some problems of operating the personnel function in the construction industry will also be briefly considered.

Importance of the personnel function

It is frequently said that people are the most important asset of any organisation. It is therefore to be expected that a recognisable personnel function, often with the label of human resources management attached, will always have a very high priority. However, this is not always the case, for a number of reasons:

1. It may be fashionable to stress the importance of people, but sometimes this is image rather than substance, and deeds do not match words.
2. Due to different management perspectives, introduced in Chapter 2, there may be different perceptions of what 'looking after your people' actually means.
3. There may be differences in how the personnel function is carried out, with the role dispersed rather than concentrated in a personnel department.

4. The importance of the personnel function may vary over time according to economic circumstances and the legal framework prevailing.

It is reasonable to assume that the tone for the personnel function will be set by top management, notwithstanding the scope for separation of ownership and control discussed in Chapter 5. Therefore, the perspectives and style of management will be influential. Where the unitary perspective is held, coupled with a paternalistic management style, then personnel may have a significant role in matters such as welfare and motivation strategies. If the style is more autocratic, then there may be little inclination to benign behaviour towards the workforce, and personnel staff may find their role limited to advising management on matters such as the minimum legal obligations with which the firm must comply. Where the pluralist perspective is held, there is likely to be a degree of collective bargaining in the organisation. In these circumstances, personnel staff could well have a major role to play in the management of industrial relations. This will be discussed in Chapter 17.

The approach to personnel management may not remain fixed over time. There may be internal changes such as new owners, directors or middle management. Some of these may bring different perspectives to the organisation, which might affect the conduct of the personnel function. There are also changing external influences to be taken into account. An important example is the economic situation, particularly the state of the labour market. When there is full employment, employees are able to obtain better conditions, and employers may be compelled to take personnel matters more seriously. In times of high unemployment, some employers will regard personnel as substantially unnecessary since wages and conditions are 'controlled' by market forces. Apart from the economic situation, there may be changes in the legal framework. As this becomes more extensive and complex, the need for regular specialist advice in this field becomes increasingly necessary. This will be considered further in Chapter 18.

Organisation of the personnel function

This is a question of who is responsible for personnel management, and in particular whether it is:

- part of every line manager's task, or
- the responsibility of a specialist personnel department.

This debate is similar to the one surrounding marketing in Chapter 9. Just as in the market-orientated firm, marketing is the responsibility of everyone, so in the people-orientated firm, every manager has a personnel responsibility.

Even so, there will nearly always be some kind of personnel or human resources management department. The size and authority of the department will depend on the importance placed on this function by top management. For example, where personnel is a high priority then there is likely to be representation at director level. Otherwise personnel may be regarded as a support function, with a less senior person in charge.

The method of working can vary between personnel departments. In some cases the staff may all be generalists, with each person able to deal with all personnel matters. In other cases individuals may specialise in particular personnel functions such as recruitment, training, industrial relations and so on. Much may depend on the structure of the organisation, as discussed in Chapter 7. If the multi-divisional structure is in use, which is increasingly the case, then each of the divisions will probably have general responsibility for their own personnel matters, with the head office possibly providing support and advice in specialist areas such as employment law.

A possible advantage of separating personnel responsibility from line management is objectivity. This is particularly the case where the firm is managed on pluralist lines with collective bargaining in place. In the event of a dispute, line management and employees may be too close to the situation, and an external party may be better equipped to help resolve the dispute. In this respect, the personnel manager and the trade union official may play similar roles. Of course, both look after the interests of those who employ them, but there is also the element of professional detachment. This is not dissimilar to the role ascribed to professional consultants, such as quantity surveyors and architects, employed by the construction client, but who nevertheless are expected to retain an independent professional role.

Having discussed the importance of the personnel function and how it is organised, particular aspects of personnel management can now be considered in more detail.

Recruitment and selection

These two processes are usually thought of as a 'matched pair', but they should be distinguished:

1. Recruitment is a personnel function whereby the firm goes out into the marketplace to attract potential employees.
2. Selection is making the choice from those who come forward. The selection process is likely to have a major input from those who will actually be working with the potential employee.

Thus, recruitment is best thought of as a marketing exercise, as the firm is trying to sell itself to the potential employees it would like to engage. This topic can be dealt with under a number of headings:

- recruitment policies
- job descriptions
- candidate specifications
- selection procedures and processes.

Recruitment policies

There are certain policies which will remain constant for reasons of efficiency, good employee relations or good practice, and others which might change according to circumstances such as the state of the labour market. Some of these policies may be explicit and appear in advertising, while others may be part of a hidden agenda and never openly stated.

To obtain a flavour of how these various factors influence recruitment policies, some examples will be given. A common policy is to operate an internal labour market — that is, all vacancies will first be advertised within the organisation. Only if this process fails to attract a satisfactory candidate will the position be advertised more widely. Some possible reasons for this policy are:

1. There may be a considerable saving on recruitment costs such as advertising, induction and training.
2. The successful applicant will be someone known to management, and presumably compliant with the organisation's procedures.
3. The practice may be part of a collective bargaining agreement, and helpful to good industrial relations.

Other recruitment policies include:

- not to discriminate in terms of gender, race, disability, religion, sexual orientation or other characteristics;
- always to inform candidates of the progress of their applications, and not leave them waiting too long for a decision;
- always to recruit under own banner rather than through an employment agency.

There are a number of policies which may be operated informally, such as a preference for recruiting older workers. This might be because they are considered more reliable, although some firms may think they will accept lower wages given the lack of alternative employment opportunities open to them.

Indeed, the labour market can have an important influence on recruitment policies. The higher the level of unemployment, the more precise a firm can be in recruitment. For example, it will be able to state its requirements more specifically in an advertisement, and in all likelihood attract suitable candidates. When there is full employment, staff may be difficult to find, and an advertisement is likely to be worded more generally in the hope that a 'wide trawl' may catch the right person.

Another problem for management is that during a recession the number of applicants for each vacancy may be very high, even if the advertisement is fairly specific. In this case various informal filtering processes may be operated such as ruling out at once anyone outside a particular age range, or anyone without a particular paper qualification, even though it may not really be essential for carrying out the job. This enables a shorter list to be drawn up for further consideration. This may be the most practical course of action, but those qualified candidates who have been rejected will feel that the firm has discriminated against them.

Job descriptions

Candidates for employment will probably be most attracted by the job description. The essential features of a job description will be included in any advertisement. More details are likely to be included in further particulars, and more again will be forthcoming at interview stage. Some obvious items that describe a job, include:

- title
- salary
- conditions such as hours, holidays and pension schemes
- content and responsibilities of job
- place in the organisation.

In some cases the firm may not be able to specify the job in all respects and there will be some scope for the successful candidate to tailor it to their own ideas.

Candidate specifications

This is complementary to the job description mentioned above. It lays down standards against which candidates can be compared, and is of great interest to the employing organisation. Although some details of this may be included in the information available to candidates, either in the advertisement or in further particulars, it is likely that much will be restricted for use as a checklist by management.

One of the best-known frameworks for the examination of candidate specifications is the Seven-Point Plan developed by Professor Alec Rodger in the 1950s, and is as follows:

1. Physical make-up
2. Attainments
3. General intelligence
4. Specialised aptitudes
5. Interests
6. Disposition
7. Circumstances.

This forms a list which organisations can adapt to their own requirements. Some aspects of the specification may be appraised from the application form, while others will need an interview before an assessment can be made.

Selection procedures and processes

Each organisation will have its own procedures for dealing with recruitment and selection. Even within the same organisation the procedure will differ according to the kind of job to be filled. Procedures can vary between a notice of vacancies outside the factory or shop, through to an elaborate multi-stage process involving advertising, short-listing from the applicants, personality tests, and two or more rounds of interviews. Furthermore, as previously stated, the condition of the labour market affects the difficulty likely to be faced in finding the right candidate. This, in turn, will affect selection procedures and processes.

Selection is a two-way process, with candidates needing the opportunity to assess employers as well as vice versa. For most jobs, some kind of written application will be required. This may be in the form of a general letter of application and/or a curriculum vitae, or it may include a standard application form. The latter has the advantage of a standardised response, which might make comparisons between applicants easier. It also makes it easier for the firm to obtain the required information. A more open application format allows the candidate to expand on, and emphasise, those aspects likely to enhance his or her application.

After applications have been received, there will be a short-listing exercise which may be carried out by the personnel department. Depending on the number of applications, this might appear quite arbitrary, with many ruled out on the basis of age or lack of paper qualification, as discussed previously. Organisations do this for practical and economic reasons because the next stage of the selection process is likely to involve interviews, and becomes more time-consuming and expensive.

Even if short-listing is the responsibility of the personnel department, at interview stage there will be involvement from line management, since they will be involved with whoever is appointed. This gives rise to one of the criticisms of interviews, which is that they are often carried out by amateurs who have had little or no training in interviewing technique. In addition, those conducting the interviews may not have had time to fully prepare. They may not have had time to read all the applications carefully, and therefore may allow themselves to be swayed by others.

Training

Training concerns building the human capital of the nation and therefore has an importance far beyond the needs of the individual organisation. Furthermore, it is not a discrete process because it relates to the associated issues of education and career development. Education, training and development have in common a concern for building human capital, but there are differences:

1. *Education* is primarily for the benefit of the recipient, and gives a range of broad knowledge and skills which will be of use in life and work generally.

2. *Training* is more specific to a given job, and is therefore of more direct benefit to the employer.
3. *Development* combines education and training — it is not so much about a specific job to be done now, but looks to the future needs of the individual and the employer.

These three processes do not take place in isolation, and there is much debate about the boundaries. In considering further and higher education, for example, some courses are clearly non-vocational and therefore can be regarded as 'pure' education. Others, such as craft courses, contain a substantial element of specific job training. There is also a grey area in deciding, for example, how to classify degree courses in the construction disciplines. These have a vocational element, but there is sometimes disagreement on the extent to which they should concentrate on the specific skills required by employers, rather than giving students a broad adaptable base which is more beneficial to them in the long term.

Since training is both costly, and a matter for public policy, the question arises: Whose responsibility it should be, and in particular, who should pay for it — the individual, the employer, or the state? Training is an investment just as much as expenditure on fixed assets, but whereas the firm owns the latter, this is not completely the case with investment in training. Contractual provisions may place some restrictions on an employee taking his or her skills elsewhere, but it cannot guarantee performance for the firm making the investment.

The problem for the firm which wants to invest in training is that it runs the risk that its trained staff will be poached by other firms, who, because they have spent little or nothing on training, can afford to pay higher wages. One possible solution is for all firms to be required by law to pay a levy towards training. Those firms who train would be able to claim funding, while those who do not train, would not as a consequence be able to obtain a competitive advantage. Schemes of this sort do exist, for example, through the Construction Industry Training Board. However, they are not as rigorous as schemes in other European countries, and training in the UK economy generally is still regarded as problematical. If anything, current government policy is leaning towards individuals paying a larger proportion of training costs themselves, the move away from student grants towards loans being one example.

Having outlined some of the underlying factors affecting the training environment, the fact remains that a successful firm will need to have a training policy in place and this is an important personnel function. The firm needs to decide:

- what type of training is required
- who is to provide it
- where it is to take place
- who is to pay for it.

Type of training

Within a firm, a wide range of people will be employed with different training requirements. The skills required by a particular type of worker may not remain constant, and could become greater or lesser. For example, it was shown in Chapter 10 that production methods have been de-skilled, thus requiring a different type of manual worker than was traditionally the case. In the construction industry, traditional apprenticeships have all but disappeared, with training now based on NVQs. These concentrate on the competences required for doing specific tasks, rather than the ability to plan and execute a range of tasks. This may not necessarily be detrimental if it reflects the changing technological requirements of the production process.

Providers of training

The traditional way to learn is by working alongside somebody already qualified. This could easily be informal, unstructured and generally ad hoc. And, of course, the results could be extremely variable. While there is a place for informality, training now tends to be more formal with the greater involvement of professional trainers, inside or outside the firm. Given the overlap with education that was referred to previously, some public provision might also be expected. The training organisation which is retained in-house will depend on the size of the firm, its range of training requirements, and company policy. For example, it might be decided, as a matter of policy, to carry few overheads and to use external consultants on an ad hoc basis, especially if most training requirements are of an occasional, specialised nature.

Place of training

Informal training will often occur at the workplace as described above. Even if this is not the recognised training method, it is quite likely that most employees will learn a good deal this way, and therefore it should be encouraged. Much training may also occur away from the workplace. This could be within the firm or at external locations such as colleges and training centres.

Payment for training

This matter has previously been discussed and it was noted that payment may be by the individual, the employer or the state, or some combination of all three. There have been changes in recent years. In accordance with market forces, especially human capital theory, the government now encourages individuals to invest in their own education and training to a larger extent. Similarly, employers seem less inclined to support staff on day release courses. Frequently students are expected to pay their own fees, and study time must come out of annual holiday, or the time must be made up by working longer hours on work days. No doubt these trends have something to do with the condition of the labour market. Whether this will give the country the investment in human capital it needs remains debatable.

Job evaluation

One of the main reasons for carrying out job evaluation is to obtain some sort of objective scientific basis for setting levels of pay for different jobs in the organisation. This is not the only way to set wages, and indeed in Chapter 17 various other ways will be considered. Even where collective bargaining is in place, job evaluation may be an agreed basis for settling differentials between jobs. Collective bargaining is still important for settling overall wage levels as well as other conditions of employment.

Job evaluation is meant to be fair and objective and focused on the job, not the individual doing it. The personnel department, being detached from production, is well placed to direct the process, given co-operation from the other departments and managers involved. Evaluation methods, at their most straightforward, may be no more than a ranking of different jobs in some kind of order. Alternatively, they could be more analytical with each job assessed against a standard list of factors which make up the 'worth' of a job. These factors could include:

- education and training required
- experience required
- originality and creativity in the job
- complexity of the job
- responsibility for others
- mental and physical demands of the job
- difficulty of the conditions under which job is done.

Each firm will need to draw up its own list, and ascribe relative importance to each factor.

Performance appraisal

This is possibly the most controversial aspect of personnel management because it concerns assessment of the individual, unlike most other aspects which are related to the job. Of course, appraisal always exists in an informal way, since managers will have opinions about the performance of their subordinates, which they may even write down.

Formal performance appraisal is more systematic and requires a considerable administrative input, probably from the personnel department. It is expected that appraisal systems will be introduced by agreement with the workforce, and with the trade unions where collective bargaining operates.

The stated purpose of appraisal systems is that they should benefit employees as well as the employer, by treating the process as an opportunity to discuss career development. The appraisal is normally based on an interview between the employee and a manager, possibly with an appraisal form having been completed as a preliminary. As a result of the interview, there should be an agreed plan of action which could include, for example, ways of improving performance, a training programme, promotion or transfer to other duties.

The appraisal will be carried out against an agreed set of criteria, and any appraisal form used should conform to these criteria. As expected, the criteria used are likely to differ between organisations, and within a given organisation, different criteria may apply according to the job. The actual criteria may emphasise the qualities of the individual, such as knowledge and reliability, or they be more results orientated, emphasising matters such as numbers of new customers attracted or external research funding obtained.

Although the stated aim of performance appraisal systems is to be objective, many employees are suspicious of them, fearing that the results may be used to their disadvantage. In particular, in times of high unemployment, many worry that appraisal will be used as a basis for wage cuts or even selection for redundancy. There is also the problem, mentioned before in the context of recruitment and selection, of the lack of experience in interviewing. Like any other skill, training is required. Personnel professionals may have difficulty persuading line managers that they should allow themselves adequate time to be trained in the skills of performance appraisal.

Other personnel functions

In this chapter some of the main personnel functions have been discussed. There are, of course, quite a few more, and in this section some of these will be mentioned very briefly, to allow for further discussion in the relevant chapters in Parts Three and Four of this book. These additional functions include responsibility for:

- collective bargaining, including negotiations with trade union representatives
- employment law, including keeping managers informed of any changes
- health and safety, including the formulation of policies and compliance with regulations
- incentive schemes, and advising on other motivation matters
- the management of change insofar as this affects employees.

As stated earlier in this chapter, personnel functions are carried out by a combination of line managers and personnel specialists, in proportions which vary between firms and over time.

A final point to be made is that although personnel management is as important in construction as in any other industry, and, in respect of matters such as health and safety, is more important than most, there are certain factors which tend to reduce the influence of personnel matters:

1. Since construction is carried out on decentralised sites, some of which are quite remote from head office, there may be difficulties in providing a consistent and thorough service.
2. Because of the preponderance of subcontracting and self-employment, most workers on site are not employees of the main contractor, thus creating ambiguities of responsibility.

Summary

This is the final chapter in Part Two, 'Functions of management', and has provided the link to Part Three, 'Human aspects of management'. The importance of the personnel function has been discussed together with the responsibility for performing it. Some of the more usual personnel functions, such as recruitment and selection, training, job evaluation, and performance appraisal, were discussed. Finally, very brief consideration was given to other personnel functions which will be examined later. This paves the way for the next part of the book which will consider some human aspects of management in more detail.

Further reading

For general management reading on personnel management, see Cole (1993), chs 45–8 and 51; Dixon (1991), ch. 10; and Graham and Bennett (1992), Part 2. For human resources management in construction, see Langford *et al.* (1995). Other reading related to human resources management generally and in construction will be given in Parts Three and Four.

Part Three

Human aspects of management

13 Management behaviour

Introduction

Until this point, the emphasis has mainly been on management as a series of processes. For example, Part Two of this book considered the various functions which have to be performed in order to manage an organisation effectively. In Part Three, the emphasis will be on the human aspects of management. In this opening chapter the focus will be on management behaviour. Whereas the previous chapters were concerned with management, this chapter is concerned with managers.

Firstly, reference will be made to Chapter 2, where the various management perspectives were introduced. Consideration will then be given to the kinds of matters which motivate managers and define their jobs. This will lead to an examination of management roles. This concentrates on what managers actually do, rather than the functions they perform, as discussed in Part Two. The work of Mintzberg is very influential here, and is widely quoted in management literature. Finally, the various skills which managers need to carry out their roles will be discussed.

There is, of course, a great deal of variety between what different managers do and the skills required. It is particularly important to realise that managers at different levels of the structure, and performing different functions, could be very different in terms of roles and skills required. One of the key aspects of management behaviour is leadership. Although this will arise in this chapter, it will be examined in more detail in Chapter 14.

Management perspectives and styles

It was established at the outset of this book that management is not an objective science, and that there is no universal view on the best way to manage. Organisations are different, and individuals involved in management bring with them a variety of intellectual positions and value systems which affect the way they manage. Among the important conclusions to be drawn are:

1. The attitude to the market may determine the attitude to the workforce, in

terms of how pay and conditions are determined. This topic will be examined in Chapter 17 on industrial relations.

2. Assumptions about what motivates people will affect the way work is organised. This, too, will be explored later, in Chapter 15.

According to the perspectives of managers on these and other matters, management styles will be adopted. These could range from autocratic and authoritarian, through a form of benign paternalism, to more democratic styles which give a good deal of autonomy to employees.

What motivates managers?

A good deal of attention is paid to employee motivation, but it is also worth considering this from the point of view of managers. For example, research described by Male and Stocks in Hillebrandt and Cannon (1989), identifies the need for:

- power
- affiliation
- achievement.

It was found that power was important to those who aspired to leadership roles in management, while affiliation is important for group dynamics, as may be found in project teams. An analysis of quantity surveyors found a high need for achievement, thought to be linked to the close involvement they have with the economics of construction projects. There are several variables at work here, depending on matters such as size of organisation and type of job. It is quite possible that the quest for power will be a strong motivator for managers while they aspire to climb the managerial ladder. This may remain, but possibly be tempered by a desire for affiliation, or the desire to gain the respect of colleagues. The meaning of the need for achievement is not always clear. It could mean achievement of personal ambitions, or it could be more group based.

Defining the manager's job

A useful description has been provided by Stewart (1983), who describes the manager's job as consisting of:

- **constraints**, including resource limitations, physical location, and attitudes and expectations of others;
- **choices**, including what work is done, and how it is done;

- **demands**, including minimum criteria of performance, and procedures that cannot be ignored.

Each managerial job will have a different mix of these three. In some cases the job is substantially about following established procedures, that is, meeting demands. Many will prefer this to having to make decisions about which path to follow — that is, the need to make choices. For others, a job which offers little scope for autonomy and making choices will be very frustrating. It is expected that the higher up the managerial hierarchy, the more scope there will be for making choices. For the organisation, a problem arises if those who have reached high levels are not comfortable with making those kinds of decisions. Similarly, it is a problem if talented people are trapped at a lower level where they are frustrated by numerous constraints. Of course, it may be part of the motivational strategy of top management to release the creative choice-making ability of middle managers, by reducing constraints and demands.

The roles of managers

The classification of roles described by Mintzberg (1973) has been very influential. This is an attempt to describe what managers actually do, and has been seen as an important contrast with Fayol's functional approach described in Chapter 8. Mintzberg has identified ten roles for the manager, which are divided among the following three groups:

- interpersonal roles
- informational roles
- decisional roles

It is not necessarily the case that all managers perform all roles. Indeed, in many organisations, particularly those not of the usual business type, Handy (1993) argues that the term 'manager' may not be used at all and that these roles are normally performed by different people. In particular, interpersonal roles are separated out and referred to as 'leading', while informational and decisional roles, referred to by Handy as 'administrating' and 'fixing' respectively, are performed by others.

Interpersonal roles

Referred to by Handy as 'leading', these include the following three roles:

1. **Figurehead**, which is normally a senior role and involves being the public face of the organisation, representing the organisation at functions, and acting as a high-level spokesperson to the media.
2. **Leader**, which is a role which could be played at all levels of the organisation, from executive directors to first line supervisors, and in fact

wherever there is a relationship between manager and subordinate.

3. **Liaison**, which involves all forms of contacts, written or verbal, with other persons or organisations, including clients, customers, suppliers and subcontractors.

Informational roles

Referred to by Handy as 'administrating', these include the following three roles:

1. **Monitor**, which is a control function, involving the collection of information on matters such as whether company budgets, or planned project cost, time and quality targets are being adhered to.
2. **Disseminator**, which involves making sure that everybody in the organisation is provided with the appropriate information and data required for them to carry out their job properly, including, of course, the information collected by the monitoring process described above.
3. **Spokesperson**, which involves both external public relations in representing the organisation to the outside world, and internal representation on behalf of a department within the organisation.

Decisional roles

Referred to by Handy as 'fixing', these include the following four roles:

1. **Entrepreneur**, which is an active role in improving the performance and position of the organisation through, for example, initiating and implementing changes such as revising organisational structures, breaking into new markets and introducing new technologies.
2. **Disturbance handler**, which involves dealing with both external disturbances such as changes in government policy, or increases in the price of resources, and internal disturbances such as personality clashes and industrial disputes.
3. **Resource allocator**, which involves deciding how the organisation's resources, including capital, will be deployed among competing uses — a process which is particularly important within the multi-divisional structured firm.
4. **Negotiator**, which involves discussions inside and outside the organisation on matters such as contracts, payments, and the transfer of information and resources.

These ten roles are by no means discrete. The work of Mintzberg has been widely analysed, and in each case there will be a slightly different interpretation of what each role should include. There are also considerable overlaps. For example, some kind of spokesperson role is appropriate within both the interpersonal and informational groups. In the interpersonal context this is likely to be a senior person representing the image of the organisation to the media. However, in the informational context this is more likely to be functional, involving the flow of practical information.

Each managerial job will have its own mix of the ten roles, but some general guidelines can be given:

1. Top management jobs are likely to have a higher interpersonal content. This may mean that more time is spent answering questions from the media, chairing strategy meetings, sitting on high-level industry-wide bodies.
2. Supervisory management will have more decisional roles, such as solving material supply problems, resolving disputes between subcontractors, and planning the use of an item of plant on hire.
3. Middle management could well have a substantial informational role, monitoring progress on projects and passing information between office and site.
4. In smaller firms, the roles are likely to be more mixed, and even top management may be heavily involved in decisional roles.
5. The role mix of a managerial job may not remain static over time, especially if organisational culture — that is, the philosophy on which the organisation is run — alters due to, say, changes in ownership or top management.

The individual manager will perform better with one mix than another. Few people are likely to be equally good whatever the balance between the ten roles. An individual may be very competent in one type of managerial job, and poor in another. This is probably the basis of the slightly tongue-in-cheek 'Peter principle' that people tend eventually to be promoted to the level of their own incompetence. Obviously, it is good for both the individual and the organisation if people are in appropriate jobs. Appraisal schemes, as discussed in Chapter 12, may help to identify what is appropriate.

It may seem that ten roles is rather a lot for anyone. At this point it is worth distinguishing between role load and work load. In the first, the emphasis is qualitative and in the second quantitative. It is quite possible to be overworked to the point of breakdown, even though few roles are being undertaken. In fact, a heavy loading in few roles could be boring and therefore bad for motivation. At the same time, too many roles, or role overload, can also be damaging as the individual may be intellectually over-stretched, or possibly trying to do too much in an unsuitable role.

In concluding this discussion on the roles of management, mention will be made of Handy's medical analogy of the manager as a General Practitioner (GP). One role not emphasised by Mintzberg, but which Handy believes underlies the other ten, is that, 'the manager, like the GP, is the first recipient of problems'. Therefore, the first thing for a manager to decide is the same as for a GP, that is, whether there is a problem or not. Therefore, the process to be followed is:

- identification of symptoms
- diagnosis of the cause of the problem

- decide on the solution, that is the treatment
- commence the treatment.

Remaining with the medical analogy, it is important to understand that symptoms are only outward signs which may mask a host of problems. Thus there is a danger that if the treatment is chosen on the basis of symptoms alone, without a proper diagnosis, then that treatment will not work. Some managers or commentators may have a favourite answer to every problem. For example, it is sometimes said that strikes are 'caused' by militant shop stewards, and their removal would resolve all industrial relations problems. This is unlikely to be a solution as the shop steward, as the workplace representative of the employees, is merely the outward sign or symptom of the strike. The problem is more likely to have been caused by a complaint about wages, conditions or work patterns. Therefore, the correct treatment can only follow if a correct diagnosis is made.

How managers spend their time

Many of the theories of management behaviour, by Mintzberg, Stewart and others mentioned, are based on surveys carried out among managers about how they spend their time. There is a certain amount of data available, which has been classified as proportion of time spent

- on different activities
- with different groups of people.

Time on activities

In the research into how chief executives spend their time, Mintzberg identified the following activities (with proportion of time shown as a percentage):

- desk work sessions (22)
- telephone calls (6)
- scheduled meetings (59)
- unscheduled meetings (10)
- tours (3).

Thus verbal communication activities, such as meetings, telephone calls and tours of workplaces take up 78 per cent of time. Although it would appear that only 22 per cent of time is spent at the desk thinking and writing, no doubt there is a good deal of thinking time outside the observed and recorded hours. The figure of approximately 80 per cent verbal communication time has been found to be consistent with other studies on management use of time, including middle managers.

Another aspect of Mintzberg's findings is the actual amount of time spent on each activity. It was found that 49 per cent of activities lasted less than 9 minutes, while only 10 per cent of activities lasted more than an hour. It is interesting to wonder whether this meant that activities were not carried out thoroughly enough, or whether an effective system of delegation was in place.

Time with groups of people

In analysing which groups chief executives spent their time with, Mintzberg's categories can be divided into:

- subordinates (48)
- non-subordinates (52).

The non-subordinates are further classified as:

- clients (3)
- suppliers and associates (17)
- peers and trade associations (11)
- directors and co-directors (12)
- others (8).

Therefore, in summary, chief executives appear to spend approximately half their time with those below them in the organisation, and half with those on the same or higher level, or outside the organisation.

Management skills

Having considered the roles of managers, and how they spend their time, another approach to understanding the manager's job is to discuss the skills required by managers. The list of skills to be examined derives from the work of Katz (1971), and are as follows:

- conceptual
- human relations
- administrative
- technical.

All managers will require some measure of all these skills, but the particular mix depends on the level of managerial position, the nature of the organisation, and the function which the manager performs. In considering these skills, comparison with the previously discussed roles can be made.

Conceptual skills

These might be thought of as the intellectual skills which managers need. For example, a manager needs the ability to understand where there is a problem, and collect information so that it might be diagnosed and solved. This is similar to the earlier section regarding the manager as a GP. Conceptual skills also embrace thinking strategically about the organisation, and planning ahead rather than simply reacting to what happens. It involves taking a view of what should be the organisation's long-term approach to matters such as possible future markets, investment in production facilities, and personnel policies.

This appears to emphasise the senior strategic aspects of management, and indeed it is expected that the ability to think in terms of the organisation as a whole is most important for senior managers. However, senior managers do need other skills — note, for example, the figurehead role discussed earlier. Similarly, less senior managers will be more effective if they have the ability to think broadly about their own areas of responsibility, and if indeed they are allowed to do so. As will be discussed in Chapter 15, there is a view that motivation is greater when employees, including those lower in the managerial hierarchy, are allowed more autonomy to work out the best way of carrying out their responsibilities.

Human relations skills

Whereas conceptual skills may be operated alone, human relations skills concern the way a manager interacts with others, whether they be peers, more senior, or less senior in the organisation, or outside the organisation. For many, the ability to get people to do what you want is the core of management. This, of course, is leadership, which will be more fully discussed in the next chapter. But apart from managing people, the other human relations skills are concerned with being able to work as part of a group, as this will also gain the respect of colleagues. This is very important in the context of project teams where the ability to get on with a range of other individual professionals will do much to secure the successful completion of the project.

A human relations skill which is much valued is the ability to communicate, in a written and/or verbal form. Written communication through clear and concise reports for example, are vital sources of information in any organisation. Rightly or wrongly, verbal skills often create an even greater impression in meetings, interviews and the like.

Perhaps the most important human relations skill of all is the ability to understand people, what makes them tick and what motivates them to work most effectively towards the objectives of the organisation.

Administrative skills

The term administration is sometimes taken to mean activities which are in some way supportive of the main purpose of the organisation, and therefore supportive of the main thrust of management. In other words, it is about putting policy into practice rather than formulating it. If so, then it is extremely important that these skills are not undervalued. It is one thing to have those exercising conceptual skills coming up with wonderful ideas about what the organisation should be doing, and those with human relations skills communicating these ideas to customers and the world at large. It is quite another thing to be able to put these ideas into practice. It is the administrator who arranges for paperwork to be done and procedures to be complied with. It is often a deputy rather than a leadership role in terms of the organisation as a whole, although, of course, in a large organisation the person in charge of administration will have a team to manage.

Technical skills

Broadly speaking, technical skills are more likely to be exercised the closer the manager is to the production process. Technical skills are the ability to put into practice the knowledge, tools and techniques of a particular sphere of activity. This means doing jobs such as laying bricks, preparing bills of quantities, or drawing up financial statements. Many managers may spend part of their time doing these things, and part of their time supervising others doing these things. The higher up the managerial hierarchy, the less likely it is that a manager will be required to exercise these skills.

Many managers have a technical background, and it is of course a matter of debate whether this is essential to be an effective supervisor. For example, many believe that to be a good site manager, a craft background is essential, whereas others believe a degree-level education is more valuable. It may depend on whether the job of site manager requires predominantly technical or conceptual skills. It is doubtful whether there is a definitive answer to this. It might depend on the size of project, and type of technology involved.

Summary of skills requirement

It is not easy to generalise about which of the four sets of skills described will be most useful to a manager. All managerial jobs will require some measure of each of the skills. Supervisory managers are likely to benefit more from technical skills as they are closer to production. But they also need good human relations skills to get the job done, because a good craft worker will not necessarily have the managerial skills which are needed to get others to do the job.

Some problems which managers face

In this last section of this chapter some of the problems which managers encounter will be briefly considered as a prelude to subsequent chapters on leadership and motivation. These are described by Handy (1993) as 'managerial dilemmas', and are as follows:

- the dilemma of the cultures
- the dilemma of time horizons
- the trust–control dilemma
- the commando leader's dilemma.

The dilemma of the cultures

It was previously mentioned that an organisation's culture is the philosophy under which it is run. It can vary over time, and between different levels or divisions of the organisation. The successful manager needs to be flexible and adaptable while at the same time remaining a thinking individual. Those who cannot adapt will find themselves restricted to those parts of the organisation where their culture predominates. Determining the type of culture is a matter of leadership and will be further discussed in the next chapter.

The dilemma of time horizons

Handy writes about the difficulty of living in two time dimensions, the present and the future. The higher a person rises in the organisation, the longer the time horizon is. However, although more time is spent thinking of the future, the responsibility for the present remains. There are similarities between this and the idea introduced in Chapter 3 which has been a constant theme in this book — firms have long-term objectives to be profitable, but in order to guarantee survival, attention must be paid to short-term liquidity.

The trust–control dilemma

When considering organisational structures in Chapter 7, one of the issues discussed was delegation. Given that the manager remains responsible for the actions of subordinates, many are reluctant to delegate in any great degree, and wish to remain in control. However, trusting subordinates by delegating is regarded by many as the best way of motivating employees to achieve better results. It would appear that successful organisations are characterised by more delegation, but the dilemma is where to start. If trust without control comes first, then the consequences of failure can be serious for the manager. Most will be inclined to retain control until a measure of success has been achieved, following which it is easier to relax control and allow the freedom of trust.

The commando leader's dilemma

The previous dilemma was about how much to trust individuals, but this dilemma concerns the degree of freedom that groups should be given, and the extent to which the organisation should rely on such groups to achieve its objectives. According to Handy, the commando leader likes clear objectives, but great freedom to decide how they will be achieved. A project team is an example of a commando group. The objective is that the project must be completed to specified cost, time and quality targets, but how this is done is up to the project manager and team. It is not believed that organisations can be made up entirely of commando groups, because there is also the need for the stability of established management patterns in some of the more ongoing administrative functions. Many managers may wish to spend part of their careers as commando leaders — that is, project managers — but may later wish to be more involved in general management concerned with the strategy of the organisation as a whole.

Summary

This chapter has focused on the manager and what he or she is and does, rather than on management as a process and its functions. Starting from a review of management perspectives and styles, this led to consideration of the factors which motivate managers. However, the main part of the chapter was an analysis of the job of the manager. This involved examining the various management roles, some of the data which indicates how they spend their time, and the skills which they employ in carrying out their work. Finally, some of the problems or dilemmas which managers face were introduced as a preparation for subsequent chapters. One important aspect of management behaviour is leadership, which will be further considered in the next chapter.

Further reading

Many management books review the main research carried out in this field, including the work of Mintzberg, Stewart, Katz and others. Recommended reading includes: Handy (1993), ch. 11; Megginson *et al.* (1992), ch. 2; Fryer (1990), ch. 2; and Male, S. and Stocks, R. in Hillebrandt and Cannon (1989), ch. 7. See also the Bibliography for source material.

14 Leadership

Introduction

This chapter picks up some of the themes contained in the last, which discussed management behaviour in terms of managers' roles, and what they actually do. One of the main roles of management, of course, is leadership, and this will be considered in this chapter. It should be noted that in some of the literature the words 'management' and 'leadership' tend to be used interchangeably. Another earlier chapter which is particularly significant is Chapter 7, 'Organisational structures', where matters such as authority, responsibility and power were introduced. These matters are, of course, central to understanding the nature of leadership.

In studying leadership, the first task is to consider the prominence of the topic. This has not always remained constant. In this context, several types of leader will be identified and various theories of leadership will then be examined. These include theories which are based on the traits or qualities of the individual leader; those which are based on the style adopted by the leader; and those which allow for differences according to the circumstances. The latter are therefore related to the contingency school of management.

Having examined the main theories of leadership, further consideration will be given to a theme introduced at the end of the previous chapter, concerning how managers resolve the 'dilemma of the cultures'. One of the main aspects of leadership, as part of the overall management role, is to determine the culture of parts of the organisation.

Nature and types of leadership

Leadership has often been regarded as a key aspect or management. However, for a good deal of the post-war period the topic has taken something of a back seat, with more of an emphasis on democracy, in work and society. Furthermore, it has been believed that 'leaders are born not made', an idea which can be traced back to the divine right of kings to rule. This notion was pushed aside in the post-war period with the growth of management education, and the belief that management skills, and indeed leadership skills, could be taught. If the importance of leadership was downgraded, it has come back into its own in recent years, as the recession has reinforced the idea of 'the right of management to manage'.

It is not easy to define exactly what leadership is. For example, a leader may be:

- someone who is a respected part of a group, and leads by example, or simply because the members of the group think a particular person would make the best leader
- someone who is 'first among equals', acting in the capacity of a chairperson
- someone who is 'high and mighty' above the group.

These matters have been introduced earlier in the book, in Chapter 7, when authority and power were discussed. These factors are often the springboard of leadership, and hence several types of leader can be identified:

1. The **charismatic** leader's authority springs from the personality of the individual, and can be found in all kinds of organisation — business, political or religious. This type of leader fits most closely to the 'born not made' model.
2. The **traditional** leader derives authority from factors such as birth, or custom and practice. This may sometimes occur where ownership and control of a business passes within a family, but is unlikely to be sustainable on its own.
3. The **appointed** leader exercises rational-legal authority which derives from the position held. This kind of leadership does permit the 'made' rather than the 'born'.

These leadership types are by no means discrete. For example, a person may be appointed to a leadership position, but be able to exercise that leadership more effectively by having a measure of charisma. In fact, personality, or traits of the leader, form the basis of one set of theories which explain leadership effectiveness.

Traits of leaders

Given the traditional view that leaders are born, then it might be expected that the early studies of leadership focused on the personal qualities of the individual leaders themselves, rather than the way they went about their tasks, or indeed what the nature of their leadership tasks were, and who they were leading. Therefore, the first attempts to explain leadership were the trait theories. If it were possible to predict what the traits of successful leaders were, then future leaders could be selected.

Of all the studies made of leaders and their traits, it was difficult to find a great deal of commonality. However, certain traits do appear more regularly among leaders, and they come as no real surprise. They include:

- intelligence
- initiative
- self-assurance
- energy
- resourcefulness.

The lack of commonality tends to imply that the same traits would not be equally effective in all leadership situations, and therefore alternative approaches to theories of leadership are necessary.

Styles of leadership

Whereas trait theories of leadership focus entirely on the personality and qualities of leaders themselves, style theories are more related to the job to be done, and those to be led. There is still the tendency with some of the style theories to derive a universal solution, that is, a leadership style which will be effective whatever the circumstances.

Leadership styles are usually classified into the following two opposites:

- **authoritarian**, or autocratic, where the leader decides what is to be done, and imposes this on subordinates
- **democratic**, or participative, where the leader allows subordinates freedom to make their own decisions.

Handy (1993) uses the terms 'structuring' and 'supportive', regarding these as less emotive.

These two styles represent idealised models, and several attempts have been made to develop a more broadly based and meaningful classification. One of the more useful of these is the continuum of leadership styles proposed by Tannenbaum and Schmidt in 1957, and widely quoted in management literature. An adaptation of this is shown in Figure 14.1, where some of the various shades of grey between authoritarian style and democratic style are shown.

The 'authoritarian' label has a harsh ring to it. Many readers will recognise a less harsh variation which is often found in family firms and sometimes even in larger firms. Leaders or managers still make all the decisions, but rather than impose them unilaterally, they will try to implement them with agreement, in a way which represents a more **paternalistic** style of leadership. This type of style is included in the four systems of management framework, devised by Likert in 1961, as follows:

- System 1 — exploitative–authoritative, which, of course, corresponds to authoritarian.
- System 2 — benevolent–authoritative, which corresponds to paternalistic.

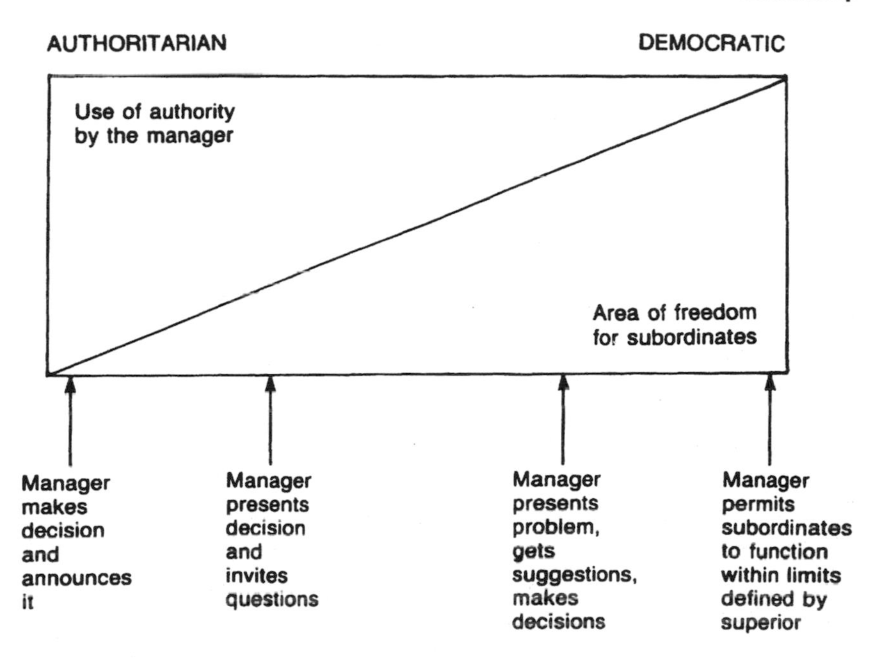

Fig. 14.1 Continuum of leadership styles (adapted from Tannenbaum and Schmidt 1958)

- System 3 — consultative, which moves towards democracy.
- System 4 — participative group, which is very much the democratic style.

It was mentioned above that many style theories attempt a universal solution; that is, they prescribe a leadership style which is appropriate in every situation. Likert's framework is an example, since he claims that system 4, the participative group or democratic style, is always superior. This is based on assumptions about human motivation. A similar argument is advanced by McGregor, writing at a similar time, in 1960. The two types of style discussed were the authoritarian Theory X, and the democratic Theory Y, with the latter claimed to be superior. The work of writers of the human relations school of management, such as Likert and McGregor, is primarily concerned with motivation, the subject of the next chapter. However, leadership style is one of the factors which helps to determine which methods of motivation are used. To continue the search for satisfactory theories of leadership, those which do not assume a universal solution must be considered.

Contingency theories of leadership

The general position of the contingency school of management is that there is no universal superior way of managing in all situations in all organisations. Therefore, translating this into the approach to leadership, contingency theories differ from trait and style theories in that they do not propose universal solutions, preferring to take account of several leadership variables.

One of the most widely discussed contingency theories of leadership comes from the work of Fiedler in the 1960s. At this time, many writers, such as Likert and McGregor described above, were arguing that democratic styles of leadership offered the best results. Fiedler's argument was that democratic styles were not always best, and sometimes more authoritarian styles were appropriate. In fact the leader should adapt the style to the circumstances.

The circumstances in question are referred to as situational dimensions, since they define the variables within which leadership is exercised. The three variables are:

1. The *position of the leader*, a major element of which is the rational-legal authority held within the organisation. This includes the jurisdiction over subordinates and the authority to implement reward-and-punishment systems.
2. The *structure of tasks* over which leadership is being exercised. This includes the extent to which the tasks are structured and defined, and the extent to which responsibility for who does them is laid down within the organisation.
3. The *relationship between the leader and the group*. This means the extent to which the members of the group like and trust the leader, including their willingness to follow his or her instructions.

If any of these variables change, it might be necessary to change the leadership style accordingly.

It was found that authoritarian styles worked best in situations which were very favourable or very unfavourable to the leader, while democratic styles worked best when conditions were moderately favourable. Whether a situation was deemed favourable depended on the three variables mentioned above, with a favourable situation existing where:

- the leader has strong authority granted by the organisation
- the tasks are clearly defined and structured
- the leader is liked and trusted by the members of the group.

An unfavourable situation is the opposite — that is, weak authority, unstructured tasks, and a lack of liking and trust between leader and group. Between these two extremes, any or all of the variables can occupy a range of intermediate positions.

Leadership is exercised at different levels of the organisation. The implication of Fiedler's work is that senior management can help individual managers to be more effective, by altering the variables to give a more favourable situation. This might include:

- enhancing the leader's authority in a well-defined way and communicating this within the organisation
- structuring more clearly the tasks to be carried out by the group as a whole, and by individuals within the group

- allocating members to groups so that a climate of trust is more likely to be fostered.

This might lead to authoritarian styles, not necessarily in the sense of harshness, but in the sense that the leader makes most of the decisions. If senior management believe a more democratic approach is better for the organisation as a whole, then this might be achieved by unstructuring the tasks while still giving the leader reasonable authority and fostering a climate of trust.

The 'best fit' approach

This approach has been proposed by Handy (1993), and can be regarded as an extension of contingency theories. Handy suggests that there are three factors which influence leadership:

- the **leader**
- the **subordinates**
- the **task**.

Furthermore, these depend to an extent on the **environment** in which leadership is exercised. Each of the three factors can be measured on a scale which runs from **tight** to **flexible**, where:

- tight is analogous with authoritarian, or 'structuring' as Handy refers to it, and
- flexible is analogous with democratic, or 'supportive' as Handy refers to it.

It is suggested that effective leadership occurs where each of the three factors 'fits' at more or less the same point on the scale between tight and flexible. If there is a lack of fit, then adjustments should be made. The three factors will be explained in a little more detail.

The leader

As stated in Chapter 2, there is no single best way to manage, and 'individuals involved in the management process bring to their work a variety of intellectual positions as well as cultural and value systems all of which affect behaviour'. The leader's value system, including matters such as attitude to employee participation, affects the preferred position on the scale from tight to flexible. Other factors affecting this include:

- the level of confidence the leader has in the ability of subordinates to control their own work
- whether the habitual style of the leader tends towards tight or flexible, since this may be difficult to alter even if it is felt to be desirable

- whether the leader regards his or her contribution to the group as vital, because if so this may tend to give a tighter style
- whether the leader requires certainty, which again will affect the degree of control, and hence tightness of desired style
- the degree of stress which the leader feels, with higher levels tending to tighter styles
- age, with older people sometimes thought to favour tight styles of leadership.

When all these factors are combined, it should be possible to place the leader somewhere on the scale from tight to flexible.

The subordinates

It cannot be assumed that people always want democratic control over their own work. No doubt some do, but some do not. The preference of a group for a particular type of leadership will be influenced by:

- confidence in their own ability
- what they understand to be the terms of the relationship between group and leader, sometimes called the psychological contract
- whether they consider their work worth being involved in
- whether they are tolerant of flexibility or prefer structure
- the extent of the experience which the group has had of working together hitherto
- cultural factors such as education, age and background of the members of the group.

The task

Whether a task should be tight, that is structured by the leader, or flexible, that is left open-ended by the leader, depends on factors such as:

- the kind of task, in terms of how routine, creative or problem-solving it is
- the time-scale of the task, since tight structures are probably faster to implement, whereas participation takes time to set up
- the complexity of the task, with technical complexity encouraging flexible styles, but organisational complexity encouraging tight styles
- whether mistakes are critical, because if they are, tighter controls may be appropriate
- the importance of the task, which may determine whether it is worth spending the time and resources in allowing flexibility.

Environmental factors

For each of the three factors considered, an appropriate point on the scale between tight and flexible can be estimated. But as mentioned previously, the environment in which these factors operate is also influential. Handy identifies six key aspects of the environment:

- the position of the leader and his or her authority in the whole organisation, as previously discussed in the section on contingency theories
- the relationship between the leader and the group, also as discussed in contingency theories
- organisational norms, or policies which place restrictions on the jurisdiction of individual leaders
- the structure and technology of the organisation which will similarly define a framework within which individual leaders operate
- the variety of tasks which have to be accommodated, with different tasks possibly requiring different leadership styles
- the variety of subordinates, which can particularly become a problem when the membership of work groups is liable to change.

These environmental factors which influence leadership are external to leader, subordinate and task but are internal to the organisation. It is suggested that a range of environmental influences external to the organisation will also affect leadership style. Some of these were discussed in Chapter 4, where the importance of matters such as market power, labour and capital markets, and the impact of public policy were considered. For example, the state of the labour market, and in particular the level of unemployment, strongly influences the style adopted by management — an issue which will be further considered in Chapter 17 on industrial relations.

In conclusion, the best-fit approach does highlight a range of complexities in determining the appropriate leadership style in a given situation. Whether leaders often think along these lines, or are prepared to adjust their style to achieve a better fit with subordinates and task is debatable It seems that many simply impose their habitual style on the subordinates and the task. This is more likely in times of high unemployment when subordinates have less power to resist, and leaders are able to impose more authoritarian styles.

Leadership and organisational culture

It will be apparent by now that it is difficult to sustain a case that there is a single theory which can be identified as the best explanation of effective leadership. In this last section of this chapter, diversity will be emphasised by taking up a theme from the section in the previous chapter entitled 'Some problems which managers face'. One of these problems was 'the dilemma of the cultures', which considers why the manager, in order to progress, needs to be flexible and adaptable, while remaining a thinking individual. This is necessary because different cultures may be present in different parts of the organisation, let alone in different organisations, or in similar organisations in different countries.

An organisation's culture may simply be described as the philosophy by which it is managed. Several cultures may be found in the same organisation, and these may change over time rather than remain static. Many of the variables discussed in this chapter will influence organisational culture, including the task to be carried out, and the value system of the leaders. Sometimes top management may lead by attempting to impose a culture throughout the whole organisation, thus giving limited autonomy to individual leaders at lower levels. In other organisations, individual leaders may be given the freedom to set their own cultures in the areas for which they are responsible.

The main types of culture identified as existing in organisations or parts of organisations are:

- the power culture
- the role culture
- the task culture.

Each will be examined, following which there will be a brief consideration of some international differences in culture.

The power culture

This is characteristic of the small entrepreneurial firm, but can also exist in larger firms. The main feature is a strong leader who sits at the centre of a web of power, and selects key individuals to maintain the power structure. As the organisation grows in size, the powerful centre may have difficulty keeping control, but this may be achieved by careful selection of individuals to control the parts. The multi-divisional or M-form of organisational structure described in Chapter 7 achieves a similar result by decentralising to other seats of power, which may well be free to adopt the culture of their choice. However, within the M-form structure, power is retained by exercising financial control from the centre.

The role culture

This is similar to the idea of bureaucracy, as discussed in Chapter 7. Bureaucracy is, of course, a very unpopular term full of negative images, but is nevertheless indicative of the culture in many organisations or parts of organisations. The term 'role culture' is less negative but means something similar, namely that the organisation is structured according to functions and tasks. The individuals are less important than the positions, and leadership is exercised not through power but through rational-legal authority. Rules and procedures are very important, and a certain lack of flexibility can be expected. Although this all sounds very uninspiring, the role culture may exist in whole organisations, especially where it is subject to limited amounts of competition. Even in more enterprising organisations it may still be appropriate to apply the role culture to those parts undertaking routine work.

The task culture

The focus of this culture is to get the job done. Therefore, it is the expected culture in successful project teams. The leader, or project manager, would see his or her job as presiding over a diverse multi-disciplinary team who respect each other and are flexible in approach. The matrix structure is sometimes used to depict project teams. Because the individual members of the team are usually highly qualified, they would expect to enjoy considerable autonomy. The leader would therefore be less involved with the work of team members on a daily basis, but will need to weld the various talents together to work towards the common goal of completing the task successfully. This can be a difficult task with such diverse people as architects, quantity surveyors, engineers and construction managers.

Some international comparisons

A far-reaching study of differences between national cultures was made by Hofstede over many years, and published in the early 1980s. The relevance to leadership comes through using the 'power distance' as one of the studied variables. Power distance measures the equality of power distribution between leader and subordinates. The higher it is, the more power leaders have in relationship to their subordinates.

The findings of Hofstede are based on surveys of employees and can be summarised as:

- countries preferring high power differences — Italy, France, Mexico, Brazil, Japan
- countries preferring smaller power differences — Denmark, Sweden, Austria, Israel, New Zealand
- countries preferring medium power distances — UK, USA, Canada, Holland.

One of the interesting features of this work is that it seeks the opinions of the led as well as the leaders. It might be expected, therefore, that any leadership style which did not conform to the national cultural preference might prove to be ineffective. However, times have changed since Hofstede carried out these investigations. As previously mentioned, employees have less bargaining power due to high unemployment, and may have to accept whatever leadership style the leaders determine. Furthermore, the growth of multi-national companies and global markets have diminished national differences. According to the findings of Hofstede, Japanese methods of management should not work in Britain. However, many Japanese firms appear to disagree, judging by the numbers who have set up operations in the UK in recent years. This was examined in Chapter 10, where it was shown how influential ideas such as just-in-time and total quality management have become in production management.

Summary

This chapter on leadership builds on the previous one on management behaviour. This is because leadership is regarded as one of the key aspects of management. Indeed, in reviewing the literature, the two words are often used interchangeably. However, this chapter sought to explain some of the distinctive features of leadership. Starting by identifying the different types of leader, this led to consideration of the main theories of leadership together with reviews of some of the writings on the subject. Thus trait, style and contingency theories were examined, followed by the 'best-fit' approach which extends contingency theory. Finally, leadership was related to cultural differences between parts of an organisation, between organisations and between countries. A theme of this chapter has been the attention paid to the led as well as the leaders. This implies that part of the secret of effective leadership is the ability to obtain good performance from those led. This is the province of motivation, which is the subject of the next chapter.

Further reading

This chapter has considered some examples of the many writings within the various theories of leadership. Source material is quoted in the Bibliography, but further reviews can be found in Handy (1993), chs 4 and 7; Dixon (1991), ch. 5; Cole (1993), ch. 11 and Fryer (1990), ch. 4.

15 Individual motivation

Introduction

Motivation is one of the most widely discussed topics in management literature. Work is, of course, often a collective or group activity, and this aspect will be considered in subsequent chapters. However, it is appropriate to start with the individual. This has been the focus for many who have written on motivation.

In examining the nature of motivation, consideration must be given initially to models of motivation which attempt to explain the process as a whole. For example, it could be argued that people have certain needs, or may be looking for certain rewards, or towards certain goals. In order to achieve these, they direct their efforts towards certain types of behaviour. As with other ideas discussed in this book, many believe that it is possible to develop universal theories of motivation which can be applied to all. However, others disagree, and instead believe that each person is different, and differentiated management approaches are therefore required.

It is important to consider the various assumptions made about people. These range from the idea that people are motivated purely by economic factors, through to the idea that people are motivated by broader matters such as fulfilment at work. Many of the theories on motivation stem from the discipline of social psychology, therefore a good deal of attention is placed on human needs as a source of motivation. Some of these will therefore be examined.

Having considered the various aspects of motivation, the concept as a whole will be revisited, using the concept of the psychological contract. At various points in the chapter, the theories discussed will be placed in the context of the current difficult and insecure economic circumstances, since in practice these can significantly influence the management of motivation.

The nature of motivation

A good deal of the material written on motivation considers the nature of motives, or human needs. As already noted, these ideas derive from the discipline of social psychology. To achieve a more comprehensive view of motivation, there is a tendency to relate motives to some desired end result. For example, Cole

(1993) states that 'the process of motivation involves choosing between alternative forms of action in order to achieve some desired end or goal'. Thus, the person is motivated to perform in a certain way, say, more productively, in the hope of achieving some target such as higher pay, increased status, or enhanced authority.

The approach of Handy (1993) utilises the motivation calculus, which, it is claimed, is unique to each individual. Again, there is an emphasis on the link between needs and results, with each individual 'calculating' how much 'E' must be invested in order to achieve the desired results. The 'E' factors in question are:

- effort
- energy
- excitement
- expenditure.

Of these, effort is the most familiar. However, it is rather a narrow concept, confined to what the individual does in the workplace. The other 'E' factors give a more qualitative dimension, indicative of the 'heart and soul' input, rather than simply the physical effort which is expended on a task. For example, each individual will 'calculate' what proportion of his or her total capacity for enthusiasm and passion should be invested in work, and what proportion should be invested in external interests, such as family or hobbies.

According to Handy, the motivation calculus is 'the mechanism by which we decide how much "E" we expend on any particular activity or sets of activities'. In understanding the nature of the calculus, the following points are important.

1. Each individual has a different calculus. For example, some individuals may be willing to invest considerable additional 'E', when faced with the prospect of enhanced rewards at work, whereas others may hold back, whatever the rewards, for fear of damaging family relationships.
2. The calculus is subjective. This follows from the previous point, and implies that the application of motivation theory is not universal.
3. The mechanics of carrying out the calculation in a given circumstance can be a combination of, on the one hand, the unconscious, or instinctive, and, on the other hand, the conscious or deliberate. Many people will tend to do what they have always done without thinking about it very much.
4. Decisions may have different time-spans, having short-term or long-term effects. This may affect the basis of the calculation.

To complete this initial section on the overall view of the motivation process, it is important to emphasise the chain of events whereby effort is perceived by the individual as leading to an enhanced performance, which, in turn, leads to rewards or results which are attractive to the individual. Of course, the individual does not operate in isolation. Within the organisation, certain constraints may exist due to factors such as management style. For managers, effective motivation means achieving the desired results. This can occur through a number of methods, including the stick and/or the carrot. To decide how to motivate their subordinates, managers must make assumptions about what motivates people. This is the subject of the next section.

Assumptions about people

As mentioned above, managers will tend to approach the problem of motivation according to their assumptions about people. It has been a theme of this book that managers bring to their work a variety of values and beliefs which affect their behaviour. In the context of motivation, two approaches will be considered regarding assumptions about people. These are:

- Schein's classification
- McGregor's Theory X and Theory Y.

Schein's classification

Schein identifies four sets of assumptions about people. These largely follow the historical development of the management schools of thought first introduced in Chapter 2. This means that as time has passed, one school of thought, with its corresponding assumptions about people, has been superseded by another. The four sets of assumptions are as follows:

- rational-economic
- social
- self-actualising
- complex.

Each will be considered in turn.

Rational-economic assumption

Although this assumption is primarily associated with the scientific management school of thought, its roots go back to the very early days of economic theory. In *The Wealth of Nations,* Adam Smith spoke of the source of economic growth in terms of everyone following his or her own enlightened — that is, rational — self-interest. In scientific management terms this means that people are primarily motivated by economic gain, and therefore the task of management is to ensure that employees are paid the appropriate rate of pay to secure effort and performance. Since this is usually taken to mean the market rate, there may be a problem of motivation when competition is high, and management feels compelled to cut wages in order to control costs. Although the rational-economic assumption is particularly associated with scientific management, this is by no means an outdated argument. Indeed, it has been the cornerstone of the monetarist philosophy which has guided UK economic policy since 1979.

Social assumption

This is primarily associated with the human relations school of thought, within which the Hawthorne experiments, conducted in the US in the 1930s, have been very influential. The essence of the assumption is that people are not primarily motivated by economic factors, but instead feel an important need to 'belong'. Thus the emphasis of management is less on the nature of the task, as is the case in scientific management, but more on the needs of people and the working of groups in the workplace. The Hawthorne experiments and work groups will be further considered in the next chapter.

Self-actualising assumption

This is associated with the post-war developments in the human relations school, led by social psychologists such as Maslow. The main argument is that people are motivated not by social needs but by self-fulfilment. Therefore, the emphasis is on matters such as the challenge of the work, and the degree of autonomy and responsibility present when carrying it out. Therefore, the role of management is to meet these needs of employees by providing interesting work. This will be considered in greater detail later in this chapter.

Complex assumption

This is the set of assumptions favoured by Schein himself, and would seem to fit with the contingency school of thought. The idea is that people are variable, and that there is no single dominant motivating factor. Therefore, there cannot be a single best managerial strategy appropriate to motivation. Not only do individuals differ one from another, but the same individual may have different needs at different times and in different circumstances. For example, it is possible to imagine that an unattached person, with no family responsibilities, may be less worried about economic factors. However, these factors may become more pressing over time if the person takes on more responsibilities. Therefore, when choosing motivational strategies, managers have to be adaptable. For Schein, the implication is that the employee and the organisation have expectations of each other. Part of this may be expressed in some legal way, such as through a contract of employment. However, some aspects of these expectations cannot be expressed in this way, and the agreement takes the form of a **psychological contract**, a concept which will be examined towards the end of this chapter, when reviewing motivation as a whole.

Although the complex assumption is widely supported, it should not be forgotten that in practice some of the earlier rational-economic scientific management views remain very influential, especially as they fit with prevailing economic and management philosophies.

McGregor's Theory X and Theory Y

This classifies the assumptions which can be made about people into two main groups. They relate to the autocratic and democratic styles of leadership discussed in Chapter 14.

Theory X

This is regarded as a rather negative set of assumptions about human behaviour. It is assumed that people are lazy, lacking in ambition, unwilling to take responsibility, resistant to change, and unable to take initiatives. The implications for management strategy are that Theory X people require strict supervision to ensure there is no slacking. Work needs to be tightly planned and controlled, and organised in great detail. There must be use of punishment-and-reward systems for matters such as pay and conditions. In short, Theory X implies the appropriateness of an autocratic management and leadership style. It might be argued that there are some similarities between Theory X and the rational-economic assumption.

Theory Y

This is regarded as a rather more positive set of assumptions about human behaviour. People have a much more active approach to their work. They are not lazy, and would like the opportunity to take responsibility for doing useful and interesting work on their own initiative, and within a dynamic changing organisation. If they do appear to exhibit Theory X characteristics, it is only because existing organisational practices have made them do so. The implications for management strategy are that Theory Y people do not require strict supervision. On the contrary, the role of management is to provide conditions where people can organise and develop their work in a way which benefits the organisation and themselves. In short, Theory Y implies the appropriateness of a democratic management and leadership style. It might be argued that there are some similarities between Theory Y and the self-actualising assumption.

The main conclusion that McGregor draws from his work is that Theory Y is a much more accurate picture of human behaviour, and therefore Theory Y management strategies should be adopted. This has similarities with Likert's patterns of leadership classification described in the previous chapter, where system 4, the participative group democratic style, is claimed to be the best approach.

Theory Y and system 4 are further examples of management ideas which propose a universal answer of best practice in all circumstances. This may have seemed reasonable when McGregor and Likert were writing, in the 1950s and 1960s, but appears less appropriate in the more difficult economic circumstances of the 1980s and 1990s, when contingency theories are more influential, and the complex assumption is perhaps more accurate.

Management approaches

A meaningful way of interpreting Theory X and Theory Y is not to say that one or the other represents the truth, and that a particular management strategy is always appropriate. Instead, it might be said that people exhibit, from the point of view of the organisation, a mixture of negative Theory X characteristics and positive Theory Y characteristics. For example, it is possible to imagine an individual simultaneously thinking:

1. It does not seem right that I should work hard to make money for the boss, and take orders from someone who shows no interest in, or respect for, my aspirations.
2. It would be good to be able to be creative in my work, and produce something I can be proud of.

If this is an accurate representation of human behaviour, then the task for management is complex. Managers need to minimise the negative aspects, and maximise the positive. It may be desirable to think that managers will concentrate on encouraging the positive, but this is not necessarily the case, since many undoubtedly adopt Theory X strategies. In many organisations the approach is mixed — the reader is referred to the core/periphery production management strategy discussed in Chapter 10. In this scenario, some workers are regarded as 'core' and will be motivated in accordance with Theory Y, while others are regarded as 'peripheral' and will be subjected to Theory X approaches such as punishment/reward systems, insecurity of employment and close supervision.

Needs

When considering the nature of motivation earlier in this chapter, it was stated that human needs form the basis of most theories. In this section, needs will be considered in more detail. There has been a good deal written on this topic in management literature, particularly derived from the discipline of social psychology, and the following will be reviewed:

- hierarchy of needs
- two-factor theory
- achievement theory
- immaturity–maturity theory.

Hierarchy of needs

This well-known concept originates from the work of Maslow in the 1950s. The hierarchy takes the form of five sets of needs, ranging from lower to higher level as follows:

- *physiological* needs, such as food, sleep and shelter
- *safety* needs, such as freedom from insecurity, physical and mental threats
- *social* needs, such as love and belonging
- *esteem* needs, such as self-respect and the respect of others
- *self-actualisation* needs, which are the highest, and include the need for self-fulfilment.

One of the features of Maslow's hierarchy is that it applies to human behaviour in general and not just to the world of work. It is argued that people need to satisfy the lower needs such as physiological, before being able to move on to their higher needs. Furthermore, it is unsatisfied needs which act as motivators. An important implication of this is that once an individual has enough money to satisfy his or her physiological and security needs, it no longer acts as a motivator to improve performance. This does not mean, of course, that people would not like more money; they will, for a whole possible range of reasons, including enhanced status.

In the work situation it is often felt that managers should implement a motivation strategy which aims to satisfy the higher levels of the hierarchy of needs. But, of course, in an economic climate where considerable unemployment and insecurity prevail, some managers may be tempted to think that fear of failure to satisfy lower physiological and security needs will be sufficient motivation for the workforce.

It is possible to simplify Maslow's five levels into three:

- physiological and safety needs, which concern well-being and simply existing
- social needs, which concern relating to other people
- esteem and self-actualisation needs, which concern personal growth and fulfilment.

Two-factor theory

This comes from the research of Herzberg and is specifically concerned with satisfaction at work. The research included interviews, and it was claimed that the findings were similar for all grades of employee. It was found that there are two sets of factors operating:

- hygiene factors
- motivators.

Hygiene factors

These factors do not act as motivators in that they will not stimulate better performance. However, they can be a source of dissatisfaction if they are not present. If they are present, they are likely to give employees a positive feeling about the organisation, and this may encourage them to want to work within it. Therefore, they act as pre-conditions, or as a foundation for motivation, but are insufficient in themselves. Hygiene factors include:

- pay
- working conditions
- interpersonal relations
- organisational policy.

Motivators

Whereas hygiene factors relate to the context or environment of work, true motivators relate to the work itself. Where these factors are also present, the employee will not only be happy to work in the organisation, but will also feel motivated to work harder to achieve better performance. These motivators include:

- interest in the work
- levels of responsibility and autonomy
- recognition and status
- sense of achievement.

The conclusions flowing from Herzberg's research are that management should direct most attention to improving job satisfaction and job enrichment, and that work should be designed to contain the maximum number of motivators. This, of course, is somewhat different to the previously dominant scientific management school, which sought to break down work into simpler tasks, over which the employee would exercise little or no control.

Achievement theory

This comes from the research of McClelland, and is notable for focusing on the differences between individuals, rather than on the common factors in motivation which is more frequently the case. Apart from physical needs, all others have been grouped together under the following categories of needs:

- affiliation
- power
- achievement.

It is thought that most people have some measure of each of these sets of needs, but that each individual has them in different proportions.

Affiliation

Most people have some need for affiliation. Where this predominates then such people are perhaps less likely to be strong decision makers, because they are likely to see securing consensus as more important than getting the task done.

Power

The need for power is one which some people can do without. However, this will not apply to those who aspire to some form of managerial position. Managers often have a strong need for power, but where this dominates it can lead to a rather authoritarian management style. Many feel that this is often not very effective, and is better when combined with a considerable need for achievement.

Achievement

The need for achievement is often seen as positive, especially if the individual's needs can be related to the needs of the organisation. Those with high achievement needs will be good at taking responsibility, and carrying tasks through. However, if this is carried too far, then the individual will be unable to operate effectively when there is a need for team working, or compliance with strictly defined organisational goals. For this reason, very high achievement needs may be effective in small decentralised organisations but less so in large bureaucratic organisations. In such organisations, the high need achiever may feel frustrated, and may possibly be in conflict with those who have greater affiliation needs.

Immaturity–maturity theory

This comes from the research of Argyris. The main argument is that people mature with time, but that the organisation fails to respond. Examples of the ways in which the personality moves to maturity from a position of immaturity are:

- becoming active rather than passive
- developing independence
- acquiring more and deeper interests
- developing longer time horizons
- developing self-awareness and self-control.

It is argued that a conventional bureaucratic organisation is based on the assumption that people are immature in terms of the above factors. For example, command structures tend to encourage people to be passive and dependent. For those who have developed some way towards maturity this is extremely frustrating. But the lesson is that demotivated behaviour is not due to laziness, which would require closer supervision. On the contrary, it is more likely to be due to too close supervision, and generally being treated as if immature.

The psychological contract

Having considered the nature of motivation, assumptions about people and a range of ideas about needs, it is appropriate to return to the concept of motivation as a whole. Using Schein's classification of assumptions, the one which fits best with current thinking is the complex assumption. It was stated that the employee and the organisation have expectations of each other, partly expressed legally through a contract of employment. It was also stated that the total expectation or 'agreement' goes beyond this and can be expressed in the form of a psychological contract which may run alongside the legal one.

There are, of course, two sides to every contract. The individual's behaviour is affected by the motivation calculus, as described earlier in the chapter. This mechanism determines the 'E' factors which an individual expends on any particular activity, or set of activities. Although the calculus differs for each individual, the other side of the contract — that is, the needs of the organisation — sets the bounds within which the contract can be negotiated.

Although the main concern here is the psychological contract at work, it should be noted that individuals do in fact normally enter into a whole range of such contracts — with family, friends, clubs and societies through which hobbies and interests are pursued, and so on. This is important, because it means that not all of an individual's needs are met, or indeed obligations discharged, through a single contract connected with work. From the management point of view, it cannot be assumed that all employees will be seeking to satisfy their higher needs, such as esteem, at work. Many will prefer to see work as the place where their lower level physiological and security needs are met, while expecting to satisfy their higher needs outside the workplace.

Contracts are not necessarily trouble-free. Problems arise where expectations do not coincide. One case may be, as suggested in the previous paragraph, where managers expect employees to be seeking to satisfy higher levels of needs than is in the minds of the employees themselves. Of course, the opposite may also be true — employees may be seeking to satisfy higher level needs at work, but find that the firm is unwilling or unable to provide work at that level. In both the above cases, there is no effective psychological contract, because the expectations of one or other of the parties are not being met.

Other problems arise where there may be an 'agreed' contract to start with, but over time the expectations of one party may change, leading the other to feel that the 'contract' has been breached. This is a common problem in the work situation where changing economic and technological circumstances may require the contract to be renegotiated. This can be difficult enough to achieve for the legal contract of employment, let alone for the more abstract psychological contract.

Attempts can be made to categorise types of psychological contract. In some organisations one will predominate, while other organisations may use different contracts for different employees. The main types are:

1. **Coercive** contracts, where the contract is rather one-sided, with one party laying down the terms, and the other having little choice but to accept them. The disadvantaged party will only expend effort or 'E' out of fear. Although this type of contract is thought to exist only in organisations such as prisons, it could be argued that the fear of being made unemployed can have a similar effect when few alternative job opportunities exist.
2. **Calculative** contracts, where the contract is of a voluntary nature. Both parties see very clearly what they want, possibly a sum of money for a certain amount of work. From the motivation point of view, all expenditure of 'E' must be negotiated and paid for, including when an increase is required. This type of psychological contract will most closely match a legal contract.
3. **Co-operative** contracts, where the contract is most flexible. The employee is likely to identify more closely with the aims of the organisation, and

likely to be more involved in the setting of those aims. Under this type of contract, the employee will expend whatever 'E' is necessary to get the job done, possibly well beyond any legally contracted hours. This is often referred to as goodwill, and can be eroded when management attempts to unilaterally change other aspects of the contract.

Summary

As stated at the beginning of this chapter, motivation is one of the most widely discussed topics in management literature. Firstly, it was appropriate to derive an outline view of the motivation process as a whole, whereby effort in pursuit of needs can be seen to lead to some desired results. However, to understand this properly, there was a more detailed examination of certain aspects of motivation. This included the different types of assumptions which can be made about people, and also the nature of the needs of the individual. It was shown that the individual's motivation calculus does not exist in isolation, but is developed within the context of the organisation and its goals. Thus there is a psychological contract between the individual and the organisation whereby terms are laid down which match the expectations of both parties. This contract exists in addition to any legally binding contract. The nature and problems associated with the psychological contract were then considered. This chapter has focused on the individual aspects of motivation. However, work is often a collective or group activity. The next chapter will begin to consider these aspects.

Further reading

The Bibliography contains reference to some of the source material by writers mentioned in this chapter, and influential in the extensive literature on motivation. For more general overviews, see Handy (1993), ch. 2; Cole (1993), chs 7 and 9; and Dixon (1991), ch. 6.

16 Work groups

Introduction

The previous chapter focused on the individual. However, as work is often a collective or group activity, it is important to establish why groups are necessary. Given that the organisation has a structure, one reason is that work groups will be needed to carry out various tasks or sets of tasks. Groups are of different types — formal or informal; primary or secondary. The construction industry is characterised by multi-disciplinary project teams.

The most well-known early studies of work groups were the Hawthorne experiments, conducted in the US in the 1930s. This research has been very influential, since it emphasised the social aspects of work. From this, many studies of aspects of groups have emerged; for example, group development, group cohesiveness, group effectiveness, and competition between groups. A good deal of attention has been directed towards teams and team roles, and it is particularly appropriate to consider this in the light of project teams in construction.

Types of group

One of the most obvious reasons why a need for groups exists is that organisations of all sizes have a structure. This was considered in some detail in Chapter 7, 'Organisational structures'. It was shown that a major problem is to set up a structure so that the different tasks can be accomplished, while retaining an integrated approach to the objectives of the organisation as a whole. That is, there is a simultaneous need for differentiation and integration. It was noted that the activities of the organisation can be grouped in a number of ways, for example, by:

- function
- product
- geography
- customer
- capital
- project.

The role of groups is a key factor in ensuring that these activities are carried out effectively.

There are several ways of analysing groups. Two useful classifications are:

- formal and informal groups
- primary and secondary groups.

Formal and informal groups

The formal group is what would be expected within the context of the organisational structure. So, if the organisation groups its activities around product types, then each product would have a group formally assigned to it. The word 'formal' does not refer to the way in which a group carries out its work, but rather to its recognised place in the organisation. A 'formal' group may in fact have an informal method of working in terms of matters such as democratic leadership style and flexibility of tasks. To overcome this problem of identification, it may be clearer to refer to 'official' rather than 'formal' groups. An informal, or unofficial group is one which the employees have created to satisfy their own needs. This will, of course, not be recognised in the formal organisational structure. This does not mean that management is unaware of such groups; indeed, it may encourage them if it means that work is carried out more effectively. It was shown in the previous chapter, under the social assumption, that being part of a group, official or unofficial, can be an effective motivator.

Primary and secondary groups

This is a slightly different way of classifying groups which emphasises size and intimacy. In an organisation, an individual may be part of a primary group and a secondary group. In the primary group, the members work closely together sharing a common purpose, not only in terms of achieving work targets, but also with regard to matters such as looking after each other's health, safety and welfare. A gang of operatives on a construction site is a very good example. This closeness may extend beyond the workplace into the wider community. For example, this has traditionally been the case in the mining industry. Groups of construction operatives working on sites away from home are another example. Secondary groups are more remote. The individual will be aware of the existence of secondary groups, but will feel less bound by them and probably less loyal to them. Examples are the construction project or the firm. An operative may feel intense loyalty to the immediate gang, but may feel less for the project in general, let alone the firm as a whole.

Groups, of course, are very common in the construction industry, existing at several levels, including the firm and the project. Multi-disciplinary project teams are of great importance, and will be considered in more detail later in the chapter.

The Hawthorne studies

Having identified different types of groups, it is important to describe briefly the research which paved the way for numerous other studies on group behaviour. The Hawthorne studies are usually regarded as the starting point of the human relations school of management, since, for the first time, a significant study of production focused on the worker rather than the work itself. This was a break with the previously dominant scientific management school, where the emphasis was on the technological rather than the social aspects of production. Scientific management tended to operate under the assumption, embodied in neo-classical economics, of homogeneous labour. By contrast, human relations recognised the non-homogeneous aspects of labour.

The Hawthorne studies ran approximately from the mid-1920s to the mid-1930s. They were initially conducted in-house, but were eventually conducted by a team from Harvard under the direction of Elton Mayo. The interesting discoveries of the in-house studies were really accidental. They had been investigating the effect of lighting on productivity, by noting how different groups responded under different lighting levels. It was found that all groups being investigated increased their productivity. When Mayo was called in, the study focused on a group of female workers assembling telephone equipment. These workers were separated from the rest and subjected to a range of physical conditions of work. It was found that output increased whether conditions were made better or worse. The conclusion drawn was that output increased because the women perceived themselves to be a special group. The studies were then extended to a group of male workers engaged in wiring and soldering equipment. By now, it was clear to the researchers that productivity had more to do with social conditions than the nature of the work. This became the focus of attention.

Some of the main findings of the Hawthorne studies can be summarised as follows:

1. Individual workers are not necessarily motivated by money or by good physical conditions, but more by the need to belong to a group.
2. A group which receives attention and is treated as 'special' will perform better than one which is not.
3. Group norms could apply on matters such as the rate of work, and all members would be expected to work at a rate which did not vary from this norm significantly.
4. Informal or unofficial groups would operate within the more formal or official groupings. This could act as a means of mutual protection in the event of perceived unfairness in company practice.
5. Managers need to be aware of social needs, so that these informal groupings will work in harmony with the formal requirements of the organisation.

Whatever their limitations, the Hawthorne studies certainly opened up alternative ways of thinking about work groups, some of which will be considered in the remainder of the chapter.

Group development

Studies have been made about how groups are formed, how they develop, establish their norms, and retain their cohesion. The study by Tuckman (1965) saw groups as moving through *four* stages of development:

1. **Forming** — creating the group through gathering information and resources. Information will be about tasks to be carried out by the group, and the methods of working.
2. **Storming** — conflicts are likely to arise out of the forming process.
3. **Norming** — after the conflict is settled, co-operation develops and group norms are established.
4. **Performing** — teamwork has been achieved, and a system is in place for effective performance.

The importance of norms, such as the setting of work rates, was indicated in the previous section, on the Hawthorne studies. Group norms depend on the individuals who comprise the group, as well as factors external to the group such as company policy and management style. These group norms may be the norms of the informal or unofficial group. Wise management will adopt policies and a style which attempts to secure a match between the informal or unofficial norms, and the formal or official norms of the organisation.

For a group to survive it needs to maintain cohesion. This means that the members of the group retain a common interest in 'sticking together' so as to retain control over their work situation, maintain rewards for all group members, and resist threats from outside the group. This may work in stable conditions; however, changes may occur due to the group being required to undertake more or different tasks. Group membership may also change due to retirement or change of employment. Therefore, over a period of time, the ability to attract appropriate new members is an important factor in maintaining group cohesion.

Group effectiveness

Group effectiveness may be defined from two points of view:

1. The official organisation view is that an effective group is one which meets production targets in terms of output, efficiency and costs.
2. The group members' view is that an effective group is one where the needs

of the group and its members are satisfied, whether it be in terms of conditions of work, social interaction, or some other criteria.

To reinforce a point already made, wise management will try to harmonise these two points of view so that, in seeking to satisfy their own needs, the group will also be satisfying those of the organisation.

A description of the likely characteristics of effective and ineffective groups has been given by McGregor (1960). Effective groups are characterised by:

- informal, relaxed atmosphere
- participative and relevant discussions
- understanding and commitment to group tasks and objectives
- members listening to each other
- conflict not ignored but dealt with constructively
- decisions reached by general consensus
- free and open expression of ideas
- tendency for leadership to be shared
- self-examination of group progress and behaviour.

The characteristics of ineffective groups are taken to be the opposite, for example, tense atmosphere, discussions dominated by a few, and so on.

This list tends to assume that the position remains fairly constant for a given group. However, it can also be argued that groups can change their effectiveness by altering certain factors. Some of these factors may be reasonably stable in the short term, and can only be changed over a significant period, while others are capable of change in a much shorter time span.

Factors which cannot easily be changed in the short term are:

- group size
- nature of the tasks to be carried out by the group
- composition of membership of the group
- environmental factors such as factory or office layouts
- organisational culture.

Group effectiveness will depend on the extent of the match between expectations of group members, and the shape of the above factors.

Factors which can be more easily changed include:

- group motivation, including the individual's satisfaction with their work, and with group membership
- group interaction, which depends on leadership, rules and procedures.

These factors are thought to be more changeable since they relate to actual behaviour in a group. Even if the factors which cannot easily be changed in the short term are unfavourable, a well-motivated group can still be effective. So, for example, in an organisational culture which places restrictions on the autonomy of employees, a well-motivated group can still achieve a great deal. This is in spite of, not because of, the circumstances. It is still worth trying to get unfavourable factors changed over a period of time, so that group effectiveness can be all the greater.

Competition between groups

Earlier in this chapter, it was restated that a problem of organisational structure is to ensure that groups are formed to carry out the various activities of the organisation, while retaining integration so that the objectives of the organisation as a whole are also met. Therefore it is necessary to consider not only what happens within groups, but also intergroup relations.

Studies have investigated what occurs when rival groups are set up. There can be advantages and disadvantages. One advantage is that people may work together more closely if there is an external threat. A disadvantage could be that competition between groups dissipates energy, so that people lose sight of the overall objectives of the organisation. However, many believe that a certain amount of competition and conflict can be creative and healthy.

The task for management is to try to encourage the benefits of competition while reducing the disadvantages. Suggestions to help achieve these aims have included:

- reward groups on the basis of their contribution to overall organisational objectives, assuming this can be measured
- base rewards to some extent on the degree of intergroup collaboration
- encourage interchange of members between groups
- prevent any competition which results in a winner-takes-all situation, since this will totally demotivate the losers.

The balance between the advantages and disadvantages of competition between groups is a fine one to achieve, and some will no doubt advocate putting co-operation rather than competition at the top of the agenda.

Teams

A 'team' does not sound very different to a 'group', and indeed there is considerable overlap. However, whereas a group can be any collection of people working together, a team implies a group of people with different but complementary skills. The team may be set up on a permanent basis, or perhaps for a single project. The team do not necessarily all work in the same geographical location, but each member contributes something towards achieving an overall result. Sometimes the members of a team may not all be employed by the same organisation.

By reflecting on the previous paragraph it can be seen how this definition often applies to a project team responsible for a construction project:

1. A range of complementary skills, including design, economic and management, will be required.
2. The team may be set up to deal with a single project, or a series of projects, possibly for the same client.
3. Some members of the team may be site-based, others office-based.
4. Team members may be employed by a number of organisations, such as the client, consultant practices, main contractor, subcontractors.

The project team will be considered further in due course.

The members of a team must be carefully selected. This is a particular problem for a project manager assembling a team from scratch. Clearly technical and professional competence is an important requirement. However, no matter how competent an individual is, little will be achieved if that person cannot function as part of a team. For example, every team member must be able to listen to advice as well as give it. Each must also see his or her own work as contributing to the whole rather than existing in isolation.

Apart from differences in function between team members, there are a number of team roles which have been identified by Belbin (1981). These are as follows:

- **Chair** This is a leadership role which involves defining objectives, understanding and properly utilising the talents of the various team members.
- **Shaper** This is also a leadership role, but is based on action towards shaping the behaviour of the team. This is leading by example rather than co-ordinating a team of equals.
- **Innovator** This is the creative thinker who suggests the good ideas. The innovator tends to think in terms of major issues rather than practical details, and is likely to be less aware of the needs of other group members.
- **Monitor/evaluator** This is a calming influence who can look at issues objectively, and possibly prevent the team moving too quickly along a potentially inappropriate path.
- **Company worker** This is a very practical role, involving turning the ideas of others into working solutions.
- **Team worker** This role is essential for group cohesion, since it involves support to other team members and encourages communication between them.
- **Resource investigator** This is an important link with those external to the team. It involves acquiring information and resources from outside which will be useful to the team.
- **Completer** This involves concentrating on getting the job done by checking on progress and maintaining a sense of urgency in the team.

The team builder would have to make use of whatever people were available. However, chair and shaper can be thought of as alternative approaches to team leadership, so it is expected that only one of these would be present. An effective team would probably only have one each of innovator and monitor/evaluator, but could have one or more of each of the other roles.

Project teams

To complete this chapter, consideration will be given to project teams particularly as they relate to construction. The idea of project teams is not new in construction. There have always been teams of people in consultant and contractor offices who have handled projects such as the preparation of a bill of quantities, the production of a complete series of working drawings, the preparation of an estimate for tender purposes. Sometimes these teams will exist for a single project, while some will be constituted on a semi-permanent basis, for example, to undertake all the work for a particular client.

However, when one thinks of the project team of today, it is usually the team set up by or on behalf of the client to run the project from inception to completion. Traditionally, the client would appoint the architect as team leader. The architect would then have overall control of the project, and would recommend the appointment of other professionals for particular purposes. However, the emphasis at present is rather more on the multi-disciplinary team of more or less professional equals, often with a significant management role for the contractor, and with the whole team presided over more directly by the client or project manager.

There are several reasons why these changes have occurred, including the following:

1. Projects are increasingly being initiated by professional clients and developers, rather than the 'one-off' client. Such clients normally have sufficient expertise to take an active part in managing the project.
2. As buildings have become larger in scale, and technically and contractually more complex, so the need has arisen for involvement by a wider range of technical specialists. As a result, it has become increasingly difficult for a design specialist, such as an architect, to control the whole process.
3. As the size and scale of projects has increased, so a wider range of public policy issues become important. These include a variety of planning and environmental matters, which require a yet wider range of disciplines to be involved in the project team.

In addition to the above points, Fryer (1990) has discussed some other reasons for giving special attention to teamwork in construction. These are:

- **Location** Several members of the team are based away from the site, and meetings between them may be infrequent. Many feel that advances in information technology and communications have made this less of a problem.
- **Different firms** Team members may be employed by different firms, each of which may have a different organisational culture. Thus the team member may have to operate in two different cultures, and may experience a conflict of loyalty.
- **Individual differences** This highlights the differences between the professions involved in construction — the type of people each profession attracts, and their patterns of education and training.
- **Late involvement** Not all team members are appointed at the outset, and even if they are, they may not be fully involved till much later. This can be a particular problem if key decisions have already been taken which affect the work of team members, but over which they had no control.
- **Teambuilding** The difficulties of putting together a project team are increased, because individual members are often only involved part time. This is because they may also be contributing to other projects. This could even mean that project leaders may not be able to engage their first choice team members.
- **Delegation** When a team is made up of a number of highly qualified professionals, it is important that they have a high degree of autonomy or they will become considerably demotivated, and the team as a whole less effective. Therefore the ability of the project manager to delegate, even though this may be to team members outside the parent organisation, becomes essential.

These reasons, and the others previously given, highlight the problems of setting up project teams in construction. In the past, differences between the professions have also been an important factor. However, more recently there has been some growth in multi-disciplinary education and training which may improve matters over time.

Summary

This chapter has moved the emphasis from the individual to the group. Starting with a description of types of groups in the workplace, the chapter considered some of the ideas on group behaviour such as group development, group effectiveness, and competition between groups. As a prerequisite for this, it was appropriate to outline the Hawthorne studies, which are regarded as marking the beginning of the human relations school of management. These studies were a springboard for a good deal of research on groups at work, as well as the social assumption of motivation as discussed in the previous chapter. The latter part of the chapter was concerned with teams, and particularly with multi-disciplinary project teams in the construction industry. Much of the material in this chapter has been based on work carried out in the discipline of social psychology. In the next chapter, the subject matter remains with the collective rather than the individual, but the context will be the industrial relations system, which utilises, among others, the disciplines of economics and sociology.

Further reading

In addition to source material detailed in the Bibliography, more general overviews of the material appropriate to this chapter are given in Cole (1993), chs 8 and 12; Fryer (1990), chs 7 and 15; and Handy (1993), ch. 6.

17 Industrial relations

Introduction

The common view of industrial relations is that it is essentially concerned with relationships between management and trade unions. In the post-war period until 1979 this was generally the case, but since then there has been a considerable change, with much greater emphasis on the individual and the market, rather than the collective. Hence greater attention is paid to matters such as managerial prerogatives and the organisation of work, matters which have arisen in previous chapters. This has been reflected by changes in the dominant management perspectives which were first discussed in Chapter 2. The pluralist and Keynesian perspectives of the post-war period have been replaced by unitary and monetarist ones.

However, just as monetarist free market economic policies are sometimes criticised, so the move away from formal industrial relations systems and collective bargaining is also now being questioned. For example, Woollacott (1995) describes how US commentators are starting to recognise that bargaining between strong firms and strong unions overcomes conflict, and is a better path to economic success than encouraging the free market to suppress wages and conditions of citizens to the lowest possible level. Woollacott claims that the policies of the US administration are beginning to reflect these ideas.

In construction, collective bargaining has rarely had very much influence, with little enthusiasm being shown by management, and union membership being traditionally low. The period of greatest pluralist influence, in the 1970s, witnessed an ever-quickening trend, in construction, towards casualisation of the workforce, and labour only subcontracting (LOSC). In some ways these characteristics of the 'flexible labour market' have been emulated throughout much of UK industry generally, although there has been disagreement on whether this has led to a sustained increase in productivity. It is certainly true that formal conflict, as expressed through strikes, has diminished. However, conflict can be expressed in many ways, including disputes between contractors and self-employed subcontractors. Arguably, fragmented bargaining of this kind can also diminish performance and there may be a need to reconsider the issue. For example, in discussing the consequences of the sporadic, unstructured nature of disputes in construction, Langford *et al.* (1995) state that 'the productivity gains of the 1980s may be tenuous and easily reversed, and a strongly centralized industrial relations system may once again be seen as desirable'.

The first part of this chapter will reiterate the main aspects of the industrial relations perspectives. Given the importance attached to the labour market in current thinking, this will be discussed in greater detail. Following this, consideration will be given to other industrial relations topics such as the participants in industrial relations — trade unions and employers; collective bargaining; industrial conflict. Finally, the role of the state will be considered as a prelude to employment law, which will be the subject of the next chapter.

Industrial relations perspectives

These were first referred to in Chapter 2, in parallel with the three main schools of economic thought. They represent alternative frameworks of thought which may be held by managers, employees or government bodies. The perspectives are:

- unitary
- pluralist
- radical.

The main points of each will be reviewed.

Unitary This is a free market approach, echoing monetarism, and is therefore characterised by an individual rather than a collective approach to industrial relations. The task of management is to pursue the profit maximisation objectives of shareholders, and if this is fulfilled, then employees will benefit. Within the unitary approach, there may be a paternalistic management style which treats employees as 'part of the family'. There might be some encouragement for the ownership of shares, but probably not much for participation in management, and certainly not for any substantial role for collective bargaining. Conflict is generally regarded as illegitimate within the unitary perspective. However, the management style will sometimes be more authoritarian, thus having something in common with the more conflictual radical approach. The unitary perspective is probably held by a majority of senior managers, and accepted by many employees. Much of government policy since 1979 has reflected this perspective.

Pluralist This is a mixed economy approach, echoing Keynesianism, and is therefore more collective in approach. The task of management is to satisfy a range of stakeholders' interests including those of employees as well as shareholders. The existence of conflict is recognised as inevitable, but of a type which is resolvable if good practice is followed. Therefore, collective bargaining is of great importance in the pluralist perspective. For many, the logical extension of collective bargaining is wider employee participation in managing the firm, that is, greater industrial democracy. The pluralist perspective is probably held by many managers, particularly those engaged in personnel work, and by the majority of trade union officials and employees. For much of the post-war period, until 1979, pluralism had a significant influence on government policy.

Radical

The radical perspective is an altogether more conflictual approach to industrial relations — a 'them and us' situation. This may be regarded as class based, because deep-rooted conflict with 'the other side' is seen as the dominant feature of the workplace. Moreover, this conflict cannot be resolved through the kinds of procedures which pluralists advocate. The areas of conflict extend beyond the usual matters of pay and conditions, to include the nature and control of work, managerial prerogatives and so on. For managers who adopt this perspective, employees and particularly trade unions should not be allowed any autonomy, and should be treated very harshly if necessary. For employees holding this perspective, the employer's interests are always in opposition to theirs, and no resolution to this conflict is possible within the capitalist system. This perspective may be held by some managers and employees, and possibly a small minority of trade union officials and leaders. There is some argument that this perspective has been evident in some government policies since 1979, since weakening the power of trade unions has been a stated objective. These policies have been pursued using more neutral terminology such as 'flexible labour markets'.

The labour market

In economic, social and political debate a good deal of attention is focused on the labour market. It has a major impact, some would say *the* major impact on industrial relations. All markets have the functions of setting prices and allocating resources, and the labour market is no exception. Also in common with other markets, the labour market is influenced by supply and demand factors, and is judged by particular performance outcomes. The labour market could be the subject of a large study on its own, and some of the main points affecting industrial relations will be considered here.

Supply factors

These determine availability in terms of quantity and quality, and are dependent on several factors.

Participation rates

In a macro sense, this means the number of people who are regarded as part of the labour force, and might be affected by matters such as school leaving and retirement ages. One of the factors which has increased the size of the UK labour force in the post-war period is the growing trend for married women to go out to work. Participation also has an industrial/occupational dimension, because whatever may be happening in the labour market as a whole, particular industries may find themselves with a deficit or surplus of workers. For example, it is often felt that after each recession many workers leave the construction industry, never to return.

Hours worked

In theory, workers can decide the number of hours they wish to work and choose appropriate jobs. This may be true to a limited extent, but undoubtedly the number of hours that the employer is willing to offer is of more significance. Research from the Centre for Economic Performance at the London School of Economics quoted in Hutton (1995a) claims that the percentage of workers in full-time tenured employment has fallen from 55.5 in 1975 to 35.9 in 1993. This is an effect of the 'flexible labour market', and although this development may suit some people, it is questionable whether this would be the choice of the majority.

Education and training

This is one of the main determinants of the quality of the workforce. It is not the case that all workers are the same, as they have different skills. This point can be considered in terms of the skills of the workforce generally, or skills available to particular industries, or in certain locations. For example, one could focus on the adequacy of craft training in the construction trades. Investment in education and training comes from the public sector, from firms and from individual workers themselves. It is always a matter of debate which option is the most effective. In line with the free market philosophy, current government policy tends to favour individual investment.

Motivation

Motivation to work, including matters such as inclination and attitude, is the other factor which determines the quality of the workforce. It has been shown earlier, for example in Chapter 15, that motivation varies between workers. In addition, management makes different assumptions about human behaviour.

Demand factors

Whatever the supply of labour may be, there are certain points relating to employer demand for labour. These may be briefly summarised as follows:

- demand for the product which the firm makes
- possibility of substitution between labour and other resources
- relative prices of labour and possible substitutes
- employer preferences for types of worker.

Again, these factors have macro and micro aspects.

Labour market performance

Unemployment implies excess supply of labour, and therefore a lack of equilibrium in the labour market. For many, the level of unemployment is the principal determinant of the industrial relations system, because of its fundamental influence on the balance of bargaining power. Thus, in times of full employment, workers obtain a measure of bargaining equality which is not present in times of high unemployment. Of course, it is a matter of great political debate as to what the 'correct' balance of power should be.

The other great indicator of labour market performance is wage levels. The simple answer to the question of wage determination is that 'it is all about supply and demand', but, of course, the issue is more complicated than this. It is certainly true that high unemployment suppresses wages, but other factors which determine wages also exist. These will now be discussed.

Comparability

This used to be a bargaining tool of trade unions, where reference was made to other agreements already negotiated in the current wage round. At present it is more likely to be used by senior executives, claiming that they must be paid as much as counterparts elsewhere, including abroad, otherwise they are likely to be 'lost' to the industry or the country.

Differentials

It is claimed that these should be maintained between groups on the basis of factors such as:

- level of skill
- place in managerial hierarchy
- level of education and training
- custom and practice.

Preventing erosion of differentials, therefore, is another favoured bargaining counter.

Ability to pay

The level of profitability of a firm may be used as a bargaining counter:

- if high, by employees to raise wages
- if low, by employers to restrict them.

Policy of the firm

It is not necessarily the case that all employers will seek to pay the lowest market wage possible. Many assume that labour is not homogeneous, and seek out what they regard as better workers for whom they expect to pay more.

Government policy

Many policies affect wages. Some work indirectly through general economic policies, while others have a more direct impact on wages. The latter include incomes policy and equal pay legislation.

Productivity
The link between pay and productivity is well established. In some industries, such as construction, a worker's entire payment will depend on productivity, particularly in the case of self-employment and labour only subcontracting. Collective bargaining often takes place on the basis of productivity, with wages increased in exchange for more productive working practices. In addition, productivity gains also derive from investment. A case is made in the article previously referred to, Hutton (1995a), that high levels of investment set up a virtuous circle of high productivity, leading to high wages, leading to more demand, leading to more employment.

Collective bargaining

As a contrast to individual contracts and individual bargaining on the basis of market forces, collective bargaining is a system which enables the terms of contracts of employment to be set collectively. Bargaining takes place between groups of workers, normally but not always through trade unions, and employers or groups of employers in employers' associations. In the post-war period, until 1979, increasing numbers of employees were covered by collective bargaining. This was the period of Keynesian demand management, when the pluralist perspective of industrial relations was at its most influential, and collective bargaining became the norm. Since 1979 the picture has significantly changed with fewer employees covered by collective bargaining and more emphasis on the individual. Nevertheless, collective bargaining still plays an important part in industrial relations. The participants, nature, levels and scope of collective bargaining will now be considered.

Participants in collective bargaining

The two main groups of participants in collective bargaining are:

- trade unions
- employers or employers' associations.

Trade unions
Membership has fallen in recent years as fewer workers are covered by collective bargaining. Traditionally, there are three types of union:

- **Craft**, where membership is based on a particular skill gained after a period of training or apprenticeship.
- **Industrial**, where membership is based on a particular industry.

- **General**, where membership is broadly based, normally across several industries, and concentrating mainly on less skilled workers.

In the face of pressure on unions to survive, there has been a considerable reduction in their number. Some have merged with others, and in many industries there is a trend towards fewer, more general, unions. In the construction industry, union membership is fairly low. An industrial union, the Union of Construction and Allied Trades and Technicians (UCATT) exists, and some construction workers are members of general unions such as the Transport and General Workers Union (TGWU), or General, Municipal and Boilermakers Union (GMB). In the event of an inter-union dispute over which union should recruit a particular group of workers, the national body, the Trade Union Congress (TUC), will normally try to resolve it.

General unions have always derived their strength from the size of membership. Craft unions, many of whom have had difficulty surviving alone, derive strength from restricting entry, a characteristic similar to professional institutions. A general indicator of union strength is density, which is a measure of actual membership, as a proportion of potential membership in a given industry, firm or workplace. But it is important to distinguish between union membership and recognition for collective bargaining. Some employees may have the former, although their union has lost, or has never had, bargaining rights.

Each union has its own structure and set of rules, and all unions tend to rely on a mix of paid and unpaid, or lay, officials. Workplace representatives are often referred to as shop stewards and frequently carry a major bargaining responsibility.

Employers or employers' associations

Some employers bargain alone, but many are members of an employers' association, most of which are based on an industry. In construction there are several, including the Building Employers Confederation (BEC), the Federation of Master Builders (FMB), the Federation of Civil Engineering Contractors (FCEC). They, too, are subject to rationalisation. For example, many national contractors who are members of both BEC and FCEC have called for closer links or even a merger.

Employers' associations are also trade associations, and are therefore also concerned with matters other than industrial relations. Many provide a range of technical and financial services. Associations, or employers individually, may belong to the Confederation of British Industry (CBI). The CBI has a national representative role, in some respects similar to the TUC.

The nature of collective bargaining

There are broadly speaking two views on this:

1. Collective bargaining is like individual bargaining, but writ large. In other words, it is substantially concerned with the **balance of power**, with workers who bargain collectively able to obtain a better parity of

bargaining strength than if they were acting individually.

2. Collective bargaining is concerned with the balance of power, but more than this it is a form of **joint management**. It is a rule-making process where conflict is resolved and results are achieved which satisfy all the stakeholders.

The balance of power view is perhaps the more easily understandable, and is the view most commonly held. It is consistent with radical employees seeking to extend collective bargaining, and unitary and radical employers and managers seeking to restrict it. The joint management view epitomises the pluralist perspective, and is consistent with support for collective bargaining found among managers as well as trade unionists. This latter view certainly gained ground in the post-war period. By 1979, legislation was proposed for the extension of collective bargaining, up to boardroom level. This legislation was never enacted, however, and since 1979 there has been a change of direction. In more recent years, where managers are sympathetic to a joint approach, it is more likely to be outside the scope of collective bargaining.

Levels of collective bargaining

There are several levels at which collective bargaining can take place, including:

- national, industry-wide
- company-wide
- workplace level.

The practice varies between industries, and at different time periods. It could range from everything being decided at a national level, to everything being decided at workplace level, with any permutation in between. For example, in the engineering industries it was common to decide matters such as basic pay and holidays at national level, but the details of manning levels, productivity, bonus rates, etc., at workplace level. This gave major negotiating responsibilities to shop stewards and supervisors, a frequently informal arrangement which could lead to fragmentation and disputes. In the public sector, the norm until recently was for most matters to be part of a national bargain. Therefore, with the exception of items such as the London weighting, pay and conditions for public sector workers, in a given job, were uniform throughout the country. This has changed considerably in recent years, in services such as health and education. The government prefers limited awards at national level, with scope for supplementary local agreements. The general trend throughout the economy has been to move towards more localised agreements. This has made life difficult for unions, who now find they have to bargain on many fronts.

In construction, there is industry-wide collective bargaining machinery, the outcome of which is the National Working Rule Agreement. This specifies wages and conditions. However, the majority of operatives are not covered, since they tend to be self-employed or employed by subcontractors. Thus, the Agreement has little impact at site level. Union membership is low, partly because UCATT resisted self-employment for many years, and did not recruit members from among these workers. This does not mean that no collective bargaining takes place. Groups of workers still collaborate at a local level, and disputes still occur, sometimes outside any union structure.

Scope of collective bargaining

This final section is concerned with what is actually included in collective bargaining. The substantive outcomes of the bargaining process such as wages, hours, holidays, bonus rates and so on will be of great interest to employers and employees. However, bargaining on these matters takes place within a set of rules and procedures. Thus there are two types of agreements:

- procedural
- substantive.

Procedural agreements

Pluralists believe that these are the key to successful collective bargaining. Procedures should be formalised at all levels, including the workplace, and should cover matters such as bargaining rights, disputes procedures and disciplinary procedures. They should be agreed and adhered to over a long period of time. The importance attached to procedural agreements is consistent with the pluralist view that collective bargaining is capable of achieving the resolution of conflict, joint management, and the most beneficial result for all stakeholders.

Substantive agreements

These are the 'end results' of pay and conditions in which everybody is interested. They are normally renegotiated annually. The intention is that with a sound procedural agreement in place, the annual substantive bargaining should present few problems.

Industrial conflict

This is a high profile aspect of industrial relations. There is a need to consider matters such as the forms, costs and causes of conflict.

Forms of conflict

For many people, industrial conflict means strikes. However, this is unrepresentative of the real situation, because any problem which leads to a loss of productivity and performance is a sign of conflict. Those in the construction industry are well aware of this, given the number of contractual disputes, some of which have their source in industrial conflict.

Apart from strikes, conflict may take the form of industrial action short of strikes. Examples include overtime bans, working to rule, or refusing to carry out certain administrative duties. Informal types of conflict include lack of co-operation with management, and deliberately working below capacity.

Costs of conflict

The effect or cost of conflict is not easy to measure. From the worker's point of view, taking strike action rarely pays for itself, as any pay rise secured is unlikely to compensate for income lost during the strike. This is why it is often argued that rather than being a sign of excessive union strength, taking strike action is a sign of union weakness, indicating its failure to settle a dispute through its bargaining strength. From the employer's point of view, production is lost and the firm may be damaged. If a dispute results in a failure to supply customers with goods or services, then apart from any contractual problem, there is the potential long-term loss of markets.

Industrial conflict in the UK

The strike record in the UK is neither particularly good or particularly bad, compared to other countries. In the past, UK industrial relations have been characterised by large numbers of short unofficial strikes, compared to the rather larger set-piece disputes in Europe. It is extremely difficult to measure days lost through forms of conflict other than strikes, but even so this must be put into perspective. Work days lost because of illness and injury have nearly always far outweighed losses sustained as a result of industrial action. Perhaps greater attention to matters such as health and safety would do more to improve productivity and performance than worrying about strikes. The final point to make is that the recorded level of conflict, as measured through strike records, has undoubtedly fallen in recent years. Some take this to mean that there has been an improvement in industrial relations. Others argue that while the strike weapon has been used less frequently, other forms of conflict, more informal and harder to measure, have taken its place.

Causes of industrial conflict

There is wide disagreement on the causes of conflict, and it is in respect of this issue that the differences between the various perspectives are most pronounced.

Unitary

Under this perspective, conflict is illegitimate, because everyone should have the common interest of the firm at heart. If any conflict does arise, then it may be due to poor communications, with management 'not quite getting the message across'. Alternatively, the conflict is being caused by agitators who are stirring up trouble among a workforce who do not have any real complaint.

Pluralist

Pluralists accept the inevitability of conflict, but see it in industrial, rather than class terms. The 'militant shop steward' is seen as the symptom or instrument, rather than the cause of conflict. Therefore, the conflict can be resolved as long as the correct procedures are in place; hence, the great importance attached to collective bargaining, particularly procedural agreements. The UK strike problem, of many short unofficial strikes, was seen as being resolvable by extending formal collective bargaining down to shop-floor level.

Radical

Radicals regard conflict not only as inevitable, but extensive, deep-rooted and class-based. In other words, the interests of labour and capital are always in opposition, and from labour's point of view cannot be resolved within the capitalist system. There is suspicion of pluralist forms of collective bargaining, which are seen as ways of diverting workers from the task of challenging managerial prerogatives. However, the radical perspective is not restricted to certain sections of labour. Some managers and politicians exhibit similar characteristics — 'the enemy within' is a phrase which has been used to describe workers on strike.

The role of the state in industrial relations

The state affects industrial relations in a variety of ways, through:

- economic policy
- legal policy
- third party intervention
- being an employer.

Economic policy

As previously mentioned, economic policy can significantly affect the bargaining climate, mainly due to its effect on the level of unemployment. Thus, deflationary policies designed to reduce inflation are also likely to increase unemployment, thus reducing the bargaining power of employees. More direct intervention could occur through incomes policies, which are designed to control wage levels directly. These are more likely to be applied in the public, rather than the private, sector.

Legal policy

There is a good deal of legislation affecting industrial relations. Until 1979 this was generally supportive of collective bargaining and workers' rights, a position which has since been reversed. This topic will be discussed more fully in the next chapter.

Third party intervention

The state is sometimes seen as a 'referee' in industrial relations matters, through the various bodies set up to deal with such issues. Industrial tribunals deal with individual matters such as claims for unfair dismissal. A well-known third party body is the Advisory, Conciliation and Arbitration Service (ACAS), set up in 1975 to improve industrial relations through the extension of collective bargaining, and by helping to resolve disputes. This was very much a pluralist agenda, and although ACAS still exists, it is less influential. This, too, will be considered in the next chapter.

The state as employer

Large numbers of people are employees of the state, in central and local government, the health and education services, and so on. Therefore, the government has great powers to influence employment practices in general. One of the major changes which has occurred in recent years is the move away from national bargaining in the public sector towards more locally based agreements and 'performance related pay'. Also, a good many public services have been privatised, or semi-privatised. This has resulted in large numbers of workers effectively being taken into the private sector, or into market-orientated public services.

Summary

Although the topic of industrial relations is not referred to in the 1990s as frequently as it was in the 1970s, it is still a very important one. The way various issues are viewed depends very much on the perspectives held by the participants, including employers, employees, trade unions, employers' associations and the state. Collective bargaining is a key issue in industrial relations, and was therefore considered in detail. Whereas in the 1970s the dominant pluralist perspective encouraged collective bargaining, the more unitary approaches adopted subsequently have returned to the individual and the market, thus requiring discussion of the labour market. A topic which is always of interest in the industrial relations context is conflict, particularly strikes. However, this is not the problem it is often thought to be. Finally, the role of the state in industrial relations was considered. A major influence is legal policy which is one of the aspects to be considered in the next chapter, on employment law.

Further reading

Among the many books which deal comprehensively with industrial relations are Jackson (1991) and Farnham and Pimlott (1995). General management books sometimes include a chapter — for example, Cole (1993), ch. 52. In the construction context, see Langford *et al.* (1995), particularly ch. 5.

Industrial relations and labour market news and issues are frequently covered in the national and trade press.

18 Employment law

Introduction

The legal regulation of employment has received a good deal of attention in recent years, and, like many other branches of the law, has enormous complexities. A knowledge of construction contracts and contract law in general is helpful because a good deal of employment law is founded on the contract of employment. The purpose of this chapter is to provide a basic framework of employment law so that the reader has an understanding of the ways in which the law impinges on the world of work. It should be read in conjunction with the previous chapter, on industrial relations.

The basic features of employment law will first be considered. In many respects there are similarities with other branches of law, for example, in the effect of civil and criminal law, common law and statute. However, what distinguishes employment law is that because work is usually a collective activity, there needs to be a collective aspect to the law which regulates it. This is in addition to the considerable amount of law which regulates the individual employment relationship. The distinction between individual and collective employment law is usually taken as the most convenient framework for studying the subject.

Basic features of employment law

Two ways in which law is often classified are:

- civil law and criminal law
- common law and statute law.

These distinctions will be considered in the context of employment law.

Civil/criminal law

Civil law is concerned with relationships between individual legal entities, be they persons or companies. Examples are the laws of contract and tort. Criminal law involves the state. Although employment law often has a high public profile, the majority of it is civil law, being concerned with the relationship between an employer, and an employee or group of employees. So when reference is made to laws which make it more difficult for workers to take lawful industrial action, it does not necessarily mean that trade unions are more open to prosecution. It means that employers have been given remedies, and can seek redress through the procedures of civil law, probably contract and/or tort. Therefore, it is the decision of the employer, not the government, to pursue any legal action.

Criminal law does have an impact on employment law, for example, through laws relating to the conduct of industrial disputes. A tactic frequently used in disputes is picketing, which in legal terms is a public order matter. Contempt of court is another aspect of criminal law. So, if in a civil case, a remedy such as an injunction is granted but not obeyed, then this becomes a matter of criminal law. Fines or even imprisonment are possible consequences.

Common/statute law

The main basis of employment law is the contract of employment. This developed principally through the common law, which is, of course, the accumulated decisions of judges in the courts. Similarly, laws related to collective matters have developed in the courts, with the law of industrial action, for example, operating through tort as well as contract.

Statutes are laws passed by Parliament. They take precedence over common law where there is a conflict. While there have always been some statutes affecting collective matters, such as industrial action, the contract of employment was largely a matter for the common law, until a series of statutes began to emerge from the 1960s onwards.

Some of the statutory provisions will be considered later, but at this stage it is worth charting the development of statute law since the 1960s.

1. From the Contract of Employment Act in 1963, there were a series of statutory protections granted to workers for the first time.
2. In the 1971 Industrial Relations Act (IRA), the Conservative government attempted a wholesale reform of industrial relations. The unfair dismissal provisions of this Act survived its repeal in 1974.
3. For the remainder of the 1970s, the Labour government passed various statutes which extended workers' rights, individually and collectively, in line with the dominant pluralist philosophy. These statutes included the 1974/6 Trade Union and Labour Relations Act (TULRA), and 1975/8 Employment Protection Act (EPA).
4. Since 1980, no fewer than nine statutes on employment matters have been passed by Conservative governments. Some of these have amended the provisions of TULRA and/or EPA, while others have introduced new provisions such as compulsory balloting prior to industrial action. In all cases, however, these statutes represent a rejection of pluralism and collectivism, and the embrace of a more unitary individualist approach.

It will be apparent that the most notable feature of employment is its classification into individual law and collective law. The greater part of this chapter will follow this framework.

Individual employment law

The contract of employment is the foundation of employment law, and has been developed historically by common law, and more recently by statute. But, in an employment situation the first thing to establish is whether a contract of employment exists at all. This is not always obvious, and a distinction must be drawn between

- a **contract of service**, where the worker is directly employed under a contract of employment, and
- a **contract for services**, where the worker is employed to carry out work as an independent contractor.

This is an important distinction to make for at least two reasons:

1. A contract of employment carries greater rights for the worker in terms of redundancy payments, notice, and the right not to be unfairly dismissed.
2. An employer is more likely to be held vicariously liable for torts committed by an employee than for those committed by an independent contractor.

One need look no further than the construction industry for a prime example of why this distinction has become an issue. The majority of site operatives are no longer directly employed by a contractor, but tend to be engaged on a self-employed basis. On the face of it, this makes the worker an independent contractor. However, many of these workers are engaged by the same employer for a long period of time, so to all intents and purposes they are 'employees'. When these cases are legally examined, several tests may be applied to determine whether a worker is an employee or an independent contractor:

- **The control test** This measures the extent to which the worker is told what to do and how to do it.
- **The organisation test** This measures the extent to which the worker is integrated into, and regarded as part of, the organisation.
- **The multiple test** This looks at a range of matters in addition to the two above, to ascertain whether, in all the circumstances, the worker is to all intents and purposes part of the organisation, or running his or her own business.

As with many similar legal tests, there is room for interpretation. When the issue was explored by Bingham (1994) in the journal *Building* he concluded that much would depend on whether the worker appeared to give the 'feel' of running his or her own business.

Given that a contract of employment does exist, then two major areas to be considered are:

- terms and conditions of employment contracts
- termination of employment contracts.

There are, of course, strong similarities between the study of contracts of employment, and contracts generally. In both cases, the terms and conditions are important, and can be incorporated in a number of ways. However, the termination of employment contracts is given greater attention than the failure to discharge a commercial contract. This is because there is more at stake. Few would disagree that the loss sustained by an individual if that individual loses his or her job, and hence livelihood, is greater than that sustained by a firm when a delivery of goods fails to arrive.

Terms and conditions of employment contracts

An assumption which the courts and common law tend to make is that contracts of employment, like any other contract, are freely entered into by two equal parties. However, this does not really reflect reality. It has already been seen that, for the employee, a good deal is at stake in terms of livelihood. Similarly, it is unreasonable to assert that the individual employee is in an equal bargaining position to the employer. For these reasons, terms and conditions of an employment contract are not left to *express* individual negotiation, or to what common law might have *implied*. In addition, terms will be incorporated from sources such as *collective bargaining* agreements, *statutes* and, increasingly, from *European law*.

Express terms

In general these should be the least troublesome clauses of a contract, since they appear to have been agreed by both parties. Since the Contract of Employment Act, in 1963, there has been a provision that the employee should receive written particulars of the contract. This provision has been strengthened in the most recent statute, the Trade Union Reform and Employment Rights Act, 1993 (TURER), which, in line with unitary philosophy, is designed to strengthen the individual contract rather than the collective.

Implied terms

Where express terms are insufficient or non-existent, then terms may be implied from common law, from the conduct of the parties, or from custom and practice. Custom and practice could refer to what is customary within an organisation. Alternatively, it could refer to what is customarily expected of someone holding a particular job title.

Terms incorporated through a collective bargaining agreement

Where a collective bargaining arrangement exists, this will normally be stated in the contract. The contents of the agreement will then be incorporated into the individual contract. It is not necessary for the entire procedural and substantive agreements to be written into each contract, as cross-referencing will suffice. Therefore, individual contracts are automatically updated each time the collective agreement is altered, for example, for an annual pay increase.

Terms incorporated by statute

It was previously stated that since the 1960s there have been various statutes in force which affect the individual contract of employment. Until then, employees who did not have the protection of a collective bargain had to rely entirely on the limited protection of common law. In the 1960s and 1970s, when the pluralist perspective was most influential, it was felt that there should be what came to be known as a 'floor of statutory rights', from which all employees would benefit. These rights would apply irrespective of whether a collective agreement existed, although the expectation was that increasing numbers of workers would be covered by collective bargaining as well. During this period, statutory rights were introduced and/or extended. These rights covered:

- written particulars of employment
- minimum periods of notice
- redundancy payments
- protection from unfair dismissal
- maternity pay
- protection from discrimination.

It is important to note that minimum standards were laid down for these and other rights. The expectation was that increasing numbers of workers would become part of a collective agreement, and bargaining to improve on them would continue.

As previously stated, the statutes enacted since 1980 have eroded these rights in line with the changed policy of government. A Keynesian/pluralist approach has been replaced by a monetarist/unitary approach which regards employment rights as an inflexibility in the labour market, a burden to employers, and therefore a deterrent to job creation. Many of the rights mentioned above are still in place, but apply to fewer workers, as the qualifying criteria have become more onerous. To qualify for these rights a worker now has to have been in employment for a longer period of time, and to work for more hours per week. It was shown in the previous chapter that, in the labour market of the 1990s, increasing numbers of jobs are part time and/or temporary, affording little or no statutory employment protection.

European law

There is, of course, a good deal of controversy about the influence that European Union law should have in the UK. Part of the debate is political, involving questions of sovereignty. The issue also has an economic dimension in that the present government regards most of these laws as similar in philosophy to the pluralist statutes of the 1970s, and therefore opposes them on the same grounds. Nevertheless, membership of the European Union obliges the UK to adhere to certain Directives.

A highly publicised aspect of European law is the 'Social Charter', or the 'European Charter of Fundamental Social Rights of Workers' to give its full title, drawn up in 1989. This Charter later influenced the Maastricht Treaty of 1992. Maastricht included an 'Agreement on Social Policy', or 'Social Chapter', from which the UK government has opted out. In spite of this, workers have been able to use European law to challenge some aspects of UK employment law. The effect which European law will ultimately have on individual employment law in the UK remains, however, far from resolved.

Termination of employment contracts

As previously discussed, the issue of termination is given close attention in the context of employment contracts because there is so much at stake for the employee. It is necessary to consider the mechanisms through which termination can take place. The legal reasons which allow an employer to terminate a contract — that is dismiss an employee — will then be considered. Dismissal is deemed to be unfair if none of these reasons applies.

There are a number of mechanisms through which termination can take place:

- notice
- end of fixed term contract
- constructive dismissal.

The first two are straightforward, but constructive dismissal requires some explanation. This occurs when the employee terminates his or her own contract as a result of the conduct of the employer. If the employer is shown to be at fault, then the employee can claim damages for breach of contract. Circumstances covered by constructive dismissal include those which make the employee's life intolerable, such as harassment, victimisation or unilaterally changing the terms and conditions of contract. An employee may be able to bring an action for constructive dismissal against an employer without actually terminating the contract, because the employer's conduct still amounts to a breach of contract which carries the remedy of damages.

There are a number of reasons why dismissal may be allowed, as defined in statutes. These are:

- **Capability** This covers matters such as health and competence. If circumstances change to make these factors relevant, then codes of practice require employers to take reasonable steps to provide alternative employment.
- **Conduct** This can sometimes be grounds for dismissal, where 'gross misconduct', variously defined, has occurred. For other forms of misconduct, dismissal should not occur on first offence. When it might be implemented, depends on various factors, including what is customary in the organisation.
- **Redundancy** This occurs when the need for a particular job in the organisation no longer exists, for example, due to changes in technology. Where there is a need for fewer people, then the method of selection for redundancy must be fair.

- **Breach of law** This refers to a situation where, if an employee continued to carry out his or her duties, he or she would be committing a criminal offence. For example, a driver who has lost his or her driving licence, say due to drink driving, could be fairly dismissed.
- **Some other substantial reason** This could be interpreted very widely and opens up all sorts of possibilities for dismissal. The circumstances in which this may be allowed are primarily economic. An example is organisational restructuring, implemented in response to changing market conditions which put the firm under financial pressure. This sounds similar to redundancy. Unlike redundancy, however, the employer is under no obligation to make payments to those dismissed.

In the absence of any of the above reasons, dismissal is deemed to be unfair. However, the list of allowable reasons is quite extensive, and leads many to believe that it is not particularly difficult for employers to dismiss workers should they so desire. There are mechanisms in place to prevent unfair dismissals — for example, the ACAS codes of practice and the system of industrial tribunals. These are meant to enable workers to have their claims for unfair dismissal heard quickly and inexpensively. But, in observing the record of results from industrial tribunals, where only about 10 per cent are found in favour of the employee, Cole (1993) concludes that, 'despite the apparent legal safeguards against dismissal, the employee continues to be extremely vulnerable to the unilateral termination of the employment contract by his employer'.

If an employee is able to show unfair dismissal, then the remedies for breach of contract would apply:

- compensation, that is, damages
- reinstatement, that is, specific performance.

In contract law generally, the courts are not inclined to enforce specific performance, but would enforce additional damages. In an employment contract, specific performance is even less likely to be enforced. So, if reinstatement is awarded by a tribunal, this does not mean that the worker will get his or her job back. Additional compensation is much more likely.

Collective employment law

As explained at the beginning of this chapter, it is the collective element to employment law which distinguishes it. Therefore, the main areas for study concern trade unions, and the rights to organise, bargain and take industrial action.

Historically, common law has tended to be hostile to combinations of workers, or the taking of industrial action. This is because they contradict ancient master/servant principles. Common law is more geared to dealing with contracts and relationships between individuals. Therefore, some sort of statutory framework existed for collective matters long before it became normal in the individual law.

A convenient framework for studying collective employment law is as it affects:

- trade unions and collective bargaining
- regulation of industrial conflict.

As explained earlier in the chapter, statutes in the 1970s were broadly supportive of collective activity, while those since have been more supportive of individualism.

Law affecting trade unions and collective bargaining

There are a number of fundamental issues to be considered, including the right to be a member of a trade union, and the right to engage in effective collective bargaining. These two sets of rights do not necessarily go together.

Freedom of association

This is the term which is used to embrace a range of concepts such as the right to be a member of a trade union, and also the right not to be a member. Freedom of association is embodied in various international declarations, including the aforementioned Social Charter of the European Union, and the International Labour Organisation (ILO) Convention of 1948.

For the right to be a member of a trade union to be made effective, there needs to be the right to organise and participate in trade union activities, and the right not to be disadvantaged by being a member. This includes certain rights to time off to be trained, and to take part in union activities. An employer cannot refuse to employ someone because he or she is, or is not, a member. And a union cannot unreasonably refuse membership to a worker.

There is a potential difficulty with the 1993 Act (TURER), which appears to allow employers to offer different wages and conditions to those who sign new individual contracts, as opposed to those who remain within the existing collective agreement. This has been interpreted as placing trade unionists at a disadvantage, and may be regarded as undermining the freedom of association.

Rights to recognition for collective bargaining

Having the freedom to join a trade union as an individual does not automatically confer the right to be recognised for collective bargaining. Freedom of association does not commit the employer to bargain. It used to be the case that if an employer refused to bargain then the union could either:

- accept this decision and perhaps try to persuade the employer at a later date, or
- try to change the employer's mind by the threat of, or by taking, industrial action.

In other words, there was no legal procedure which avoided the unilateral imposition of wages and conditions by employers, or the sometimes bitter struggle of an industrial dispute.

Of course, recognition requests did not always lead to a dispute and collective bargaining did increase throughout this century. The only significant statutory procedure was enacted under the Employment Protection Act 1975, when ACAS was given the role of encouraging and facilitating the setting up of new collective arrangements. This procedure was abolished in the 1980 Employment Act. ACAS no longer has the role of facilitator of collective bargaining, mainly because the policy of the government has been to discourage it.

The final point to make in this section is that collective agreements are not in themselves legally binding contracts, as there is no intention by the parties — that is, the employer and the trade union — to be legally bound. This is because the arrangements are too complex and, in any case, contracts already exist between the employer and each individual employee. What the collective agreement does, in legal terms, is to affect the terms of the individual contract. This is also an important factor in the next section, on industrial conflict.

Legal regulation of industrial conflict

Although industrial conflict often has a very public face, in legal terms it is more a private matter, and as such is primarily regulated by civil law. However, criminal law can play a part in certain circumstances. For example, a public order issue may arise due to the conduct of pickets during a strike; or contempt of court may arise if a court injunction is disobeyed. Generally speaking, however, the law of industrial conflict operates through contract and/or tort.

Contract

It may come as a surprise to many, but under the law there is no right to strike. The reason is that the employment relationship is governed by the individual contract of employment. Under such a contract, withdrawing labour is deemed to be a fundamental breach of contract, which entitles the employer to terminate the contract — that is, dismiss the employee. In practice this may not have happened very often because:

- it was expected that when the strike was over everything would return to normal
- an employer was obliged to treat all workers in the same way, so that the entire workforce would have to be sacked and replaced — quite an undertaking.

However, since the 1980s some employers have been more willing to dismiss workers, partly because recent statutes have given employers greater powers to dismiss selectively. So, there is no right to strike under the law of contract, but there is some provision under the law of tort.

Tort

The 'right to strike' is granted by what are known as negative protections, or immunities. This 'back-to-front' approach means that there are certain torts which people would normally be liable for, but if committed in particular circumstances in connection with a trade dispute, then the workers involved would be immune from action being taken against them. This complex situation is illustrated in Figure 18.1. Employer A and Employer B have a commercial contract for the supply of goods. The employees of A are in dispute and have been called out on strike by their union. The objective is to affect the business of A. One consequence may be that A is unable to fulfil the contract with B. The torts which could arise, and which could make the trade union officials liable, are:

- inducing a breach of contract of employment between A and its employees
- inducing a breach of commercial contract between A and B.

Normally these torts are unlawful, but *there is immunity from liability if acting in contemplation or furtherance of a trade dispute.*

These words have not changed much over the years, but the definition and interpretation placed on words such as 'immunity' and 'trade dispute' have changed. The various statutes introduced since 1980 have had the effect of redefining certain things to make it more difficult to take lawful industrial action. Some examples will be given.

1. Actions against trade unions themselves are now allowed which puts their funds at risk — previously it was the union official who was deemed to have induced the breach of contract.
2. To obtain immunity, certain stringent procedures must be followed prior to a strike. In particular, there must be a secret postal ballot which hinders the union's ability to take swift action.
3. The meaning of 'trade dispute' has been more narrowly defined to cover only matters such as terms and conditions, engagement and dismissal, trade union recognition.
4. Only disputes with own employer are lawful. Solidarity or anti-privatisation actions are now deemed political and therefore unlawful.
5. Similarly, any form of secondary action, even to support a dispute with own

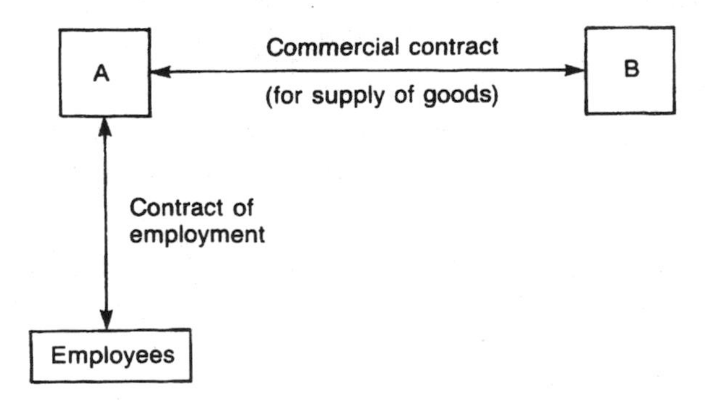

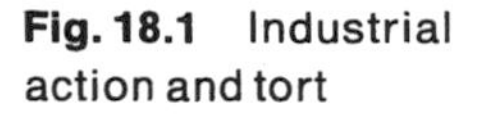
Fig. 18.1 Industrial action and tort

employer, is now unlawful. Therefore, any pressure exerted by A's employees on B, or B's employees to assist their dispute is unlawful. For example, A's employees cannot lawfully ask B's employees to boycott goods from A.

In the case of unlawful industrial action, the employer can seek **remedies**. Under the law of tort, the normal remedies are:

- **damages**, which require a full hearing to establish all the facts so that the correct level of damages can be ascertained
- **injunctions**, which prevent the tortious action continuing, and, being interim actions, require a lower burden of proof, and are thus more quickly dealt with.

It was always the case that employers were only interested in injunctions because the objective was to stop the strike and return to normal as soon as possible. There was no point in subsequently going to full hearing, for damages, because it was the trade union official who was liable for the tort, rather than the union itself.

Summary

The legal regulation of employment often has a public face, but it is mainly governed by civil law such as contract and tort. Having established this, the chapter was divided into individual and collective law. Individual employment law is dominated by the contract of employment. This is more significant to the employee than are commercial contracts between firms, because there is more at stake in terms of making a livelihood. Therefore, considerable attention was paid to the grounds on which an employment contract can be terminated as well as to the terms and conditions of the contract. The collective element is a distinctive feature of employment law, and revolves around the rights to join trade unions, engage in collective bargaining, and take industrial action. The individual contract of employment is still a significant factor, but the right to take industrial action is also strongly influenced by the law of tort.

An important theme of this chapter has been the role of statute law. Since the nineteenth century there has been some statutory regulation of collective matters, but only since the 1960s has statute played a significant part in respect of the contract of employment. The increased use of statutes implies a greater political significance. During the 1960s and 1970s, the pluralist perspective had greater influence on public policy. Therefore, employment rights and collective bargaining tended to be encouraged. But the various statutes enacted since 1980, being more unitary in character, have reversed the process so that individualism is favoured over collectivism, and it is more difficult for workers to take lawful effective industrial action. Finally, the tensions created by the more pluralist European Union law are apparent, so it will be interesting to observe how employment law develops in the coming years.

Further reading

Most industrial relations books, and some general management books, deal with employment law, including: Cole (1993), ch. 53; Farnham and Pimlott (1995), chs 9 and 10; and Jackson (1991), ch. 11. Numerous articles on employment law appear in newspapers and journals.

For an examination of the difference between employee and independent contractor in construction, see Bingham (1994). For a summary of changes to employment law through statutes since 1980, see Harper (1994).

Part Four

Management applied to the construction industry

19 Market strategies for construction firms

Introduction

Strategic management means looking into the future and deciding where the organisation wants to be. It has been shown, at various times in this text, that a major influence on the success of a firm is its competitive strength in the many markets in which it is involved, either as buyer or seller. For example, the nature of markets and competition was discussed in Chapter 4, and in Chapter 9 it was shown how firms manage their present and potential future market position in order to enhance their ability to generate revenue and profit.

For the contemporary market-orientated firm, market strategy means researching market opportunities, and putting in place appropriate financial and organisational structures. However, long-term strategy is not a free choice because the firm is subject to competitive pressures, as mentioned above, and also to short-term changes in economic conditions. As a starting point, consideration will be given to the context in which the construction firm operates — that is, the structure of the industry as it has developed, and in particular the nature of the construction market and competition. Following this the nature of construction demand and client behaviour will be considered. The construction process operates within the context of a financial system which can have a significant impact on the survival and behaviour of construction firms. Only when the foregoing points are combined, can an assessment be made of the market opportunities open to the construction firm.

Structure of the industry

The structure of many industries cannot be easily defined since they tend to have permeable boundaries. Industries consist of large numbers of firms trading with each other. The construction industry is a good example because it consists of a large variety of organisations, many of whom were defined in Chapter 3, including:

- clients
- consultants

- contractors
- subcontractors
- manufacturers of materials, components and plant
- merchants and hire companies.

All these types of organisation are required to ensure that the construction process moves through its various stages of production. This is similar to other industries, such as manufacturing, where the product evolves through raw materials, components, assembly, and sale.

In addition to defining the various types of organisation in an industry, which can be thought of as its *vertical* dimensions, it is also important to examine the *horizontal* dimensions. This refers to the number of organisations of each type, and is extremely important to the individual construction firm, because it indicates the extent of market competition it faces. As explained in Chapter 4, the extent of competition is measured by the degree of monopoly, defined as the ability to mark-up price above costs. This is determined by:

- supply factors, such as the number of firms competing
- demand factors, such as strength of demand, as measured by elasticity.

In addition, the ability of firms to collude in price fixing will also be a factor. In general, the supply factors are structural and change more slowly, whereas the demand factors are more liable to fluctuate according to economic circumstances. There is an important link, however, because in the event of a recession, and hence weak demand, smaller firms will find it more difficult to survive. This is because they tend to have fewer reserves, and less financial backing. Therefore, a long-term structural change towards greater degrees of monopoly can be set in motion by short-term fluctuations in demand.

In the construction industry, there has been a change in structure over time which has reduced competition in some respects, but increased it in others. This can best be understood by referring to the traditional structure of the industry. This mainly consisted of medium-sized general contractors who would undertake a wide range of work in their own locality, normally with a directly employed workforce, and with only specialist work being subcontracted out. The large contractors tended to specialise in civil engineering projects, and there were also some specialist housebuilders.

The picture is rather different now, as a relatively small number of large diversified national contractors have emerged. This has occurred partly through organic growth, and partly through acquisition and merger. The effect has been to reduce competition. At the same time, there has been a great increase in the amount of subcontracting, and an increase in the number of small firms competing for the work controlled by the larger main contractors. The medium-sized local or regional contractor has found it more difficult to survive on this basis, and many have gone out of business, been taken over by large firms, or retrenched into particular markets, possibly acting as subcontractors. Case studies documenting this development often appear in the construction press — see for example, Chevin (1994). The trend from the medium-sized firm, to the large and the small is not restricted to contracting. Similar trends have occurred among firms of consultants such as quantity surveyors and consulting engineers — see, for example, Barrie (1994) and Ridout (1994a).

There are a number of reasons why large firms have emerged. It is true for all industries that firms seeking to maximise profits will attempt to reduce competition by taking over or merging with others, or by colluding on price levels, and the construction industry is no exception. Some other reasons which are more specific to construction are:

- There has been a trend towards very large projects, both building and civil engineering. These projects have often required the technical and managerial abilities of large firms.
- The locational nature of construction has enabled regional monopolies to emerge.
- The production of materials and components often benefits from economies of scale and therefore lends itself to a more monopolised market structure.

While these structural trends have increased the degree of monopoly, this has to some extent been offset by international pressures:

- Increased numbers of overseas-based clients, consultants and contractors operate in the UK.
- Because building design and construction is increasingly based on high value-added components made off site (a trend to be examined in more detail in Chapter 2.3), this has made it more worth while for overseas suppliers to compete in the UK market.

To conclude this section, the effect of the recession should be reiterated. When demand falls, there is insufficient work for everyone. Therefore, supply exceeds demand, which means overcapacity in the industry. In such circumstances smaller firms tend to come under greater pressure, even if they are efficient, since they do not have the competitive and financial strength of larger firms. Calls are also made to reduce capacity by allowing firms to go out of business, or by encouraging mergers — see, for example, Cooper *et al.* (1994). This may seem appropriate in the short term, but such actions may have severe consequences in the long term if an upturn in the economy leaves the industry with insufficient capacity to meet the demands of clients.

The shape of demand

Having considered how the structure of the industry is liable to change, it is appropriate to consider the nature of demand in more detail. The construction client was discussed in Chapter 3, where it was shown that, increasingly, clients are not necessarily building for their own use. Instead they are providing space for others to use. This applies to the housebuilder hoping to sell dwellings to consumers as well as to the commercial developer hoping to let office space to organisations. It was also shown in Chapter 3 that the client role is not always vested in a single organisation. Elements, such as long-term financing, may be handled by a specialist financier who becomes part of the client body. There are client bodies who build for their own use. Often these are large retail firms with a substantial building programme who employ a range of professional expertise in-house.

The above trend, which can be called the 'professionalisation of the client', implies that clients require a greater knowledge of the needs of ultimate users, such as households or industrial and commercial organisations. What this means in terms of market opportunities for construction firms will be considered later in this chapter, but certain factors can be considered at this point. Aspects of demand include:

- quantity of demand
- quality of demand
- methods of delivering demand.

The successful construction firm needs to be aware of each of these. They will be considered separately, although in practice there is likely to be considerable overlap, especially between quantity and quality of demand.

Quantity of demand

Estimating the likely quantity of demand should be a straightforward matter of ascertaining the extent by which current building stock falls short of the required quantity. Of course, in practice the question is more complicated because buildings are:

- *long-term investments,* so demand needs to be determined well into the future
- *location specific,* so it is not just the overall shortfall in supply which needs to be determined, but its spatial distribution.

Each type of construction work needs to be considered separately, so that supply can be matched to demand over a period of time. It would appear that the industry has not been particularly good at achieving the correct balance, since speculative sectors of the market, such as housing and commercial property, have tended to be either greatly under- or over-provided, while investment in infrastructure has suffered from public expenditure restrictions.

Market research can be carried out for each of the sectors, to provide data on which assessments of future demand can be made. For housing, valuable government statistics on population trends can be studied. It is not just population as a gross figure which is significant, but the number and type of households likely to be formed. If the trend towards one-person households continues, this will affect the number of dwellings demanded and their size. In addition, there is the aforementioned question of spatial distribution — of where dwellings should be built. This is not simply a matter of consumer choice, but also a matter of planning policy. This concerns the decisions of public authorities as to where dwellings may be built and in what quantities. For a discussion on household statistics, see Barrie (1995), and on planning policies, see Moor (1995).

Quality of demand

Quality cannot always be separated from quantity, but certain considerations apply. In the case of housing, demand is generally thought to depend on effective income. This is the ability of households to spend on housing, and derives from a number of financial variables:

- actual income
- price of housing
- cost of borrowing — that is, mortgage rate taking account of tax relief.

The higher the level of effective income, the more likely it is that households can enter the market for the first time. For those already in the market, the level of effective income affects spending power. Additional spending may result in higher quantity through larger dwellings, or in higher quality. This conventional analysis of housing demand needs to be supplemented by consideration of the 'confidence' factor. Even if the conditions for effective income are favourable, the long-term investment nature of housing requires high confidence to turn this into demand. This may not be forthcoming, and it may be especially difficult to raise confidence in a slump following a boom, as was the case in the late 1980s and early 1990s.

In addition to effective income, and confidence, the other factor which affects quality of demand is tastes, including fashions. This is something the housebuilder must be particularly aware of, and indeed can take advantage of. This is because the demand for housing could apply to *all* housing. The marketing task for the construction firm is to channel this demand in the direction of *new* housing. Some of the benefits which housebuilders should emphasise were discussed in Chapter 9. It is quite likely that in difficult market conditions, potential buyers will be looking for value for money and quality in design, rather than gimmicks. However, it should be remembered that many people have a preference for a new house if at all possible, and this gives firms an advantage if they can offer an attractive package. Indications of home buyers' tastes and how construction firms are responding to these can be found in Ridout (1994b) and Stewart (1995).

Although the above discussion concerns housing, similar factors apply to other sectors of work. For example, changes in architectural fashions can affect commercial work. On matters of detail design, there was a time in the 1980s when all office buildings seemed to require air conditioning and raised floors. These together made for a considerable additional cost. In the 1990s, neither of these items is regarded as quite so essential.

Methods of delivering demand

For firms dealing in the majority of consumer goods, this may not be a significant issue. The customer may simply walk into a shop, choose the quantity and quality required, pay and walk out. For the construction firm life is not so simple. The previous sections on quantity and quality of demand concern the construction product itself, whereas this section concerns how the client obtains the product, which can be equally important.

As discussed in Chapter 8, the delivery requirements for a construction project have the following parameters:

- cost
- time
- quality.

To achieve this, the client will need to select and enter into contractual arrangements with consultants, contractors, subcontractors, suppliers and so on. This whole process is commonly referred to as **procurement**. A detailed examination of procurement is not within the scope of this book; the reader can, however, refer to a wide range of descriptions of procurement methods, including Harvey and Ashworth (1993). Nevertheless, some points can be made here.

The traditional approach to procurement is *selective tendering*, where a limited number of contractors are invited to submit a price, based on a full design detailed in drawings and bills of quantities. These are contract documents. The objective of this pure price competition is to obtain the lowest price and best value for money for the client. One of the features of this system is that the contractor is not involved in the process till quite late, and cannot contribute at the design stage. This reduces the potential contribution that the contractor could make, and market strategy is limited to trying to get onto approved tender lists. At the same time, the role of the client is also quite limited, because the contractual arrangement used gives great powers to the lead consultant, that is, the architect or engineer as appropriate.

Selective tendering still has many supporters, and is often regarded as a good solution for:

- the inexperienced client
- the publicly accountable client
- projects when speed is not essential.

However, given the trend towards the professionalisation of the client as described above, many are desirous of greater involvement. The *construction management* method of procurement, where the subcontractors carrying out the work are directly employed by the client, allows the client more involvement. It also gives contractors a significant market opportunity, because under construction management they become equal members of the professional project team. Construction management was widely used in the 1980s boom when fast completion times were required.

Another feature of the traditional system is that the separation of design and construction has meant that there is ample scope for 'buck-passing', which often results in the client paying more than expected and the project being delayed. To improve certainty of price and time, many clients have opted for a procurement method based on *single point responsibility*. This, of course, offers the contractor a very good market opportunity. The established methods in this category include package deals, and design and build.

Influence of the financial system

Having considered the supply and demand aspects of the market from the point of view of construction firms, the next problem is the availability of finance. No matter how good a market opportunity may be, it can be thwarted unless the necessary short-term and long-term finance, to support cash flow and realise profitability, is available. (Sources of finance were discussed in Chapter 6.)

- Short-term finance is working capital which enables a cash flow, and hence a programme of work, to be maintained.
- Long-term finance is required for investment, which forms the basis for sustained profitability.

For short-term finance, construction firms tend to rely on a mixture of bank credit (overdrafts) and trade credit. The latter depends on the working of the payments chain, first introduced in Chapter 3. Sources of long-term finance are usually owners' funds or borrowed funds. Traditionally construction firms grew through owners' funds being ploughed back over many years to achieve organic growth. In more recent times, firms have extended ownership beyond the entrepreneur's family by 'going public', that is, selling shares to outside investors. Also, in more recent times, construction firms have been inclined to borrow in order to finance land purchase and property development, as well as the acquisition of other firms. This has resulted in many firms becoming *highly geared,* that is, having a high proportion of borrowed funds. High gearing leaves many firms in difficulties when markets decline and/or interest rates rise, incurring high debt repayments.

Construction firms face particular problems with the financial system, particularly those whose main activity is in contracting. It was shown in Chapter 11 that contractors face greater financial difficulties than firms in, say, manufacturing and retail because they have fewer assets. Because of the ease of entry into contracting, competition is high, and this drives down profit margins. A high asset turnover, in this case a fast cash flow, is therefore required if the firm is to achieve profitability, or even to survive. Construction firms face problems in both short-term and long-term financing.

Short-term finance

For firms like contractors, who have insufficient assets to reduce risk to an acceptable level for long-term investors, great reliance is placed on short-term finance. In terms of the financial system this tends to mean bank overdrafts. These have a 'floating charge' against the firm's assets, which means that the banks have wide powers for forcing the firm into liquidation and claiming repayment of their debt against any assets the firm holds. This contrasts with long-term finance. Since this is more usually secured against specific assets, the survival of the firm as a whole is not necessarily threatened. Banks in the UK have tended not to be very sympathetic to construction firms, and have often been quick to force firms into liquidation when times become difficult. This has particularly been a problem for smaller firms and is one of the reasons why the small to medium-sized local contractor has found it difficult to survive, as described earlier in this chapter. The evidence of the 1990s seems to suggest that larger firms are retaining support from the banks, while smaller firms are finding it more difficult to do so — see Stewart (1994).

Long-term finance

As mentioned above, construction firms have tended to grow organically through private owners' funds. Of course, many firms have borrowed to a certain extent, but not too heavily in a short time period. A major exception was in the heady days of the 1980s, when funds were readily available.

There are various types of financial institutions who provide long-term finance, for example:

- banks, who lend for a rate of interest
- venture capitalists, who may lend on this basis, but are more likely to expect an equity share in the firm in exchange for finance.

Whatever the basis of the lending, it is less likely that they will be interested in the risks associated with contracting — see Dow (1995).

Market opportunities

Having considered supply factors, such as structure of the industry, demand factors in terms of quantity, quality and delivery, and the availability of finance, it is now appropriate to combine these to determine what strategies the construction firm might adopt. The key factors which have emerged are:

1. Industrial structure has tended away from predominant local- and regional-based medium-sized firms, towards a few very large diversified national firms and very many small specialist firms and subcontractors.
2. Apart from structural changes, the construction market is also affected by fluctuations in economic activity.
3. The main source of demand now derives from habitual clients, often providing space for others, rather than the one-off or occasional client.

4. The choice of methods for delivering demand has become wider.
5. Larger diversified firms have less difficulty obtaining short- and long-term finance than smaller firms specialising in contracting.

In summary, waiting to be asked to tender for building contracts does not amount to a successful market strategy, and most firms are well aware of this. In the light of the difficulties created by the severe slump of the 1990s, there is considerable debate about the future of construction firms — see, for example, articles by Birkbeck *et al.* (1995) in *Building*. This debate was stimulated by a government-funded study by Hillebrandt *et al.* in 1995. Before outlining some of the suggestions which are emerging in the mid-1990s from this very fluid situation, market strategies adopted in the recent past will be briefly considered.

As has already emerged in the course of this chapter, there has been a trend away from the medium-sized general building contractor towards the large diversified national firm on the one hand, and the smaller specialists and subcontractors on the other. The diversification process has normally embraced housebuilding, while some large specialist housebuilding firms have also emerged. There are two main routes by which construction firms have become large and diversified:

1. *Internal or organic growth,* favoured by most of the household names in construction, occurs steadily through the ploughing back of profits, supplemented by modest borrowings, but not to the extent that the firm becomes highly geared.
2. *Mergers and acquisitions,* favoured by an increasing number of firms since the 1960s, and particularly in the 1980s boom, is a faster route to growth, can be secure and successful, but does carry the risk that excessively zealous borrowing can lead to the firm being dangerously highly geared.

Some of the reasons why large firms have emerged were discussed earlier in the chapter.

The usual motivations for growth can be traced back to the desire for increasing profits. This can occur through:

- *increasing revenue* — for example, through greater market share or being involved in more sectors of the market, and/or
- *reducing costs* — for example, through greater efficiencies due to economies of scale, or the sharing and rationalisation of resources.

From the point of view of the health of the industry and the economy in general, the most beneficial method of growth is where profits are used for high-quality investment, leading to greater efficiency, economies of scale and higher income and employment. However, the nature of the financial markets often means that firms use profits to finance acquisitions. This ensures growth, market share and enhanced profits for the firm, but not necessarily greater growth for the economy generally.

The larger construction firms often had a tradition of specialising in large civil engineering projects. By diversifying, they have tended to embrace some or most of the following:

- building contracting, thus competing with medium-sized regional contractors
- housebuilding, thus competing with the larger specialist housebuilders
- property development
- construction plant hire
- construction materials
- services offering alternative methods of procurement, such as management or design and build.

This, of course, represents a list of products or services being offered to the market. Diversification also has a spatial aspect as firms have expanded from their home regions into other parts of the country and internationally — to Europe, North America, the Middle and Far East.

The debate of the mid-1990s previously mentioned, seems to suggest further diversification in general, but also some retrenchment and selectivity for the individual firm. Time alone will tell whether this represents a more balanced approach than the boom of the 1980s, when anything was apparently possible and finance was easily available. Among the developments being suggested as likely are:

- even greater divergence between large and small firms with a further 'squeeze' on medium-sized ones
- less reliance on contracting, since firms have little control over this fluctuating market
- considerable opportunities in the private provision of infrastructure, where the construction firm might provide, own and manage schemes
- link-ups between cash-flow-orientated UK firms, and asset-orientated European ones, thus facilitating the raising of capital for major projects world wide
- greater potential opportunities in overseas markets
- greater opportunities for firms to undertake work on a fee basis, including facilities management
- a willingness by many professional clients to enter into partnering arrangements — that is, long-term agreements with consultants and contractors which would make a steady work load more likely.

As mentioned above, many firms are reducing the number of markets in which they operate. This is a reaction to the 1980s boom, when many firms extended themselves too far into areas such as property development and housebuilding. In the process, many borrowed heavily and became highly geared, making themselves vulnerable to severe cash flow crises. Many firms have had to sell off potentially profitable assets just to survive. Even so, many did not survive.

Many firms who were not forced into this position have nevertheless taken a hard look at what they do and have refocused their activities. Cowe (1995) describes why one of the largest construction firms, who as part of its activities had also become the largest housebuilder in the country, now wishes to withdraw from this sector because it generates a negative cash flow pattern. The trend towards larger projects, including the possibility of a wider role for construction firms to include ownership, financing and management in their activities, suggests that medium-sized firms will indeed become even more squeezed and will have to merge, become subcontractors or develop some special niche market.

Summary

This chapter has been concerned with market strategy for a wide range of construction firms, particularly contractors. There are certain structural principles underlying this topic. In addition, variable economic conditions, which are possibly more acute for construction than for the economy generally, mean that the whole question of what the appropriate market strategy should be, is quite fluid. The first part of the chapter considered supply factors such as the structure of the industry and competition. Factors influencing demand were then considered, including the quantity and quality of demand, and the ways in which demand can be delivered or realised. Along with supply and demand, consideration was then given to the provision of short- and long-term finance before bringing these various elements together to discuss the market opportunities. Consideration was given to how this has been implemented in the past together with an outline of the debate on how this is likely to develop in the second half of the 1990s. This has implications for organisational structures, and will form the starting point for the next chapter, which will then proceed to discuss other aspects of company policy.

Further reading

This chapter has made considerable reference to earlier ones, where some of the underlying principles were discussed. Since market strategy is likely to remain fluid, many of the references for this chapter are taken from recent journal articles. Brief details were given in the text, with full details in the Bibliography.

20 Some policy matters for construction firms

Introduction

The previous chapter considered market strategies open to the construction firm, in the context of market conditions and institutions. Assuming that the contemporary firm tends to be market orientated then this requires an organisational structure which is flexible enough to respond to changing market conditions. Therefore, this chapter will commence by showing how structure matches one of the most important aspects of the firm's policy — that is, its approach to the market. Following this, consideration will be given to certain other areas of policy which affect the firm's ability to survive and grow. There are numerous areas of policy which could be examined. The ones selected are a mix of:

- those of traditional concern, such as **plant management**
- those of more recent interest, such as **quality management**
- those subject to recent changes in the legal framework, such as **health and safety**

Organisational structures

This topic was considered in some detail in Chapter 7. The general principle is that the structure of an organisation should combine:

- the ability to maintain certainty and control over the firm's activities and output, while
- taking advantage of specialisation, and remaining flexible in the market.

In considering the different ways in which a firm can group its various activities, it was concluded that the predominant structural type adopted in industries generally is the multi-divisional or M-form of organisational structure. The main features of this structure are:

1. The firm is divided into separate companies or divisions in a hierarchical fashion.
2. Each division is responsible for achieving the organisation's profitability

objectives in a given activity, which may be expressed in terms of a product, market or geographical location.

3. Each division normally has a degree of autonomy in achieving its objectives, subject to central financial control through an internal capital market.

The internal capital market involves capital being distributed to the divisions. The overall objectives are determined at head office by the group board of directors, with each level responsible for distributing capital and resources to those below. This can be a very flexible arrangement with amounts of capital being increased or reduced. Similarly, new divisions can be started from scratch, or bought, while existing divisions can be closed down, or sold off.

It should be emphasised that not all firms operate the internal capital market in exactly the same way. Some of the more financially orientated conglomerates will buy and sell virtually any company in accordance with current financial prospects, and will take little or no interest in how they are actually managed. Other firms may only choose to be involved with divisions in which they have some tradition or expertise, and will be more interested in laying down rules for how each division should be managed — for example, in terms of plant policy, quality management and health and safety practices.

It has been usual for construction firms to follow the second of the above approaches. However, in recent years conglomerates have been more involved in construction activities, especially housebuilding and materials supply. And, as described in the previous chapter, the 1980s boom witnessed many construction firms expanding quickly into all sorts of activities, such as property development, because they perceived a market opportunity to raise profits substantially.

Some general examples of M-form structures were given in Chapter 7. A more detailed example of how a national construction firm may structure itself will be given here. Such a structure is shown in Figure 20.1. In this example, there are five divisions or companies answerable to head office. Each of these is subdivided into an appropriate number of divisions. In this example it is shown that the construction division is subdivided into four divisions. Of these, the 'building' division is divided into regional companies. It is possible that individual projects may be managed at regional level, or within each region there may be more local area or district offices. For example, the South Wales Region may have area offices at Cardiff, Swansea and Pembroke. It is likely that a significant number of functions will be located at the regional headquarters. These functions include:

- estimating
- marketing
- personnel
- engineering
- purchasing
- planning
- health and safety

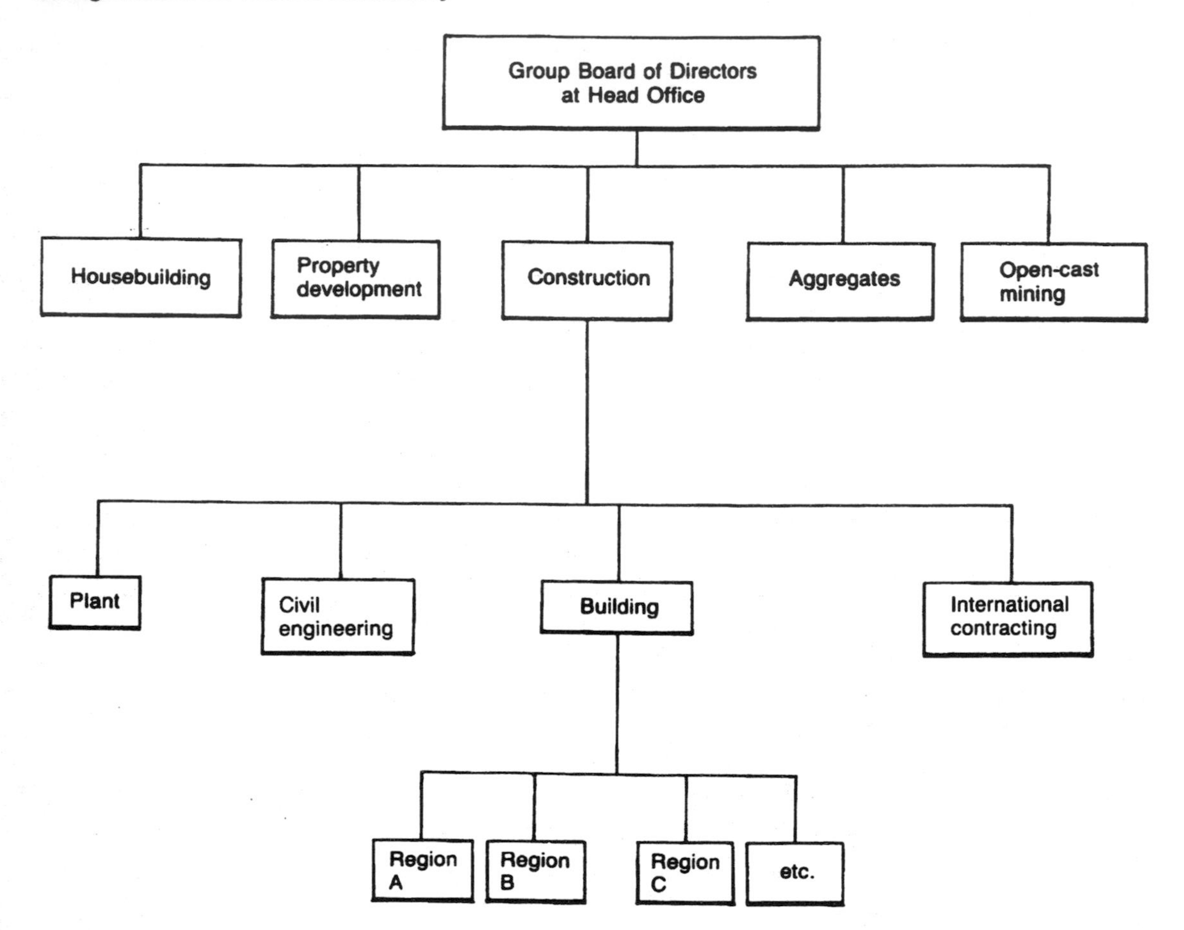

Fig. 20.1 Typical multi-divisional (M-form) structure for national construction firm

- quality assurance
- surveying
- accounting.

It is possible that some of these aspects may be decentralised to area offices, and the firm will need to formulate policies for all these functions. In some firms these will be substantially decided at regional level, while in other firms policies will be dictated from above.

Plant management

In the example above, the plant division is shown as a subsidiary of 'construction' on the same level as 'civil engineering', 'building', and 'international contracting'. In some firms, plant may be managed at regional level as a service function of the regional company. Indeed, as will be seen, deciding whether plant should be provided as a service to projects, or whether it should be organised as a separate profit-making company in its own right is one of the principal elements of a plant policy.

Another important and recurring theme regarding plant management is that any plant owned by the firm becomes a fixed cost, or overhead. This ties up part of the firm's capital and reduces its flexibility. To understand this, it is necessary to refer to Chapter 11, where the ways in which firms in different industries generate return on capital employed (ROCE) was discussed. It was shown that because high competition keeps margins low, contractors require high asset turnover to achieve adequate ROCE. The asset in question is usually cash flow, with a small amount able to generate a high level of turnover. Any capital asset, such as plant, reduces flexibility and must be fully utilised to be economic.

With these introductory points in mind, the following aspects of plant management will be further considered:

- the impact of plant usage on profit
- plant policy variables between firms
- plant implications of expansive market strategies
- whether plant should be bought or hired
- the status of a plant subsidiary company.

Plant and profits

Ultimately, the key issue is whether a given policy will help the firm achieve its objective of improving profit. Since profits derive from costs and revenue, it is worth examining the effect plant usage has on these two variables.

Costs

From the production management point of view, the use of appropriately selected plant is designed to increase productivity, and therefore reduce costs. Beyond this the impact of plant on costs depends on whether the plant is owned or hired.

1. Owned plant is a fixed cost, which will be economic if the plant is fully utilised.
2. Hired plant is a variable cost, and will be more expensive than fully utilised owned plant because the profit margin of the plant hirer must be included.

Revenue

Plant usage can help to expand markets and revenue. For example, if efficient plant usage leads to reduced costs as stated above, then this gives the firm the opportunity to offer more competitive prices and, hence, expand its market share for greater long-term profits. Of course, firms in this position often use reduced costs as an opportunity to increase profits in the short term. Nevertheless, the option of taking a more long-term view is present. Another possible impact of plant usage is that it may afford faster completions. This may be of interest to an additional group of clients looking for speed, and willing to pay a price premium.

This represents not so much an increase in market share, as an expansion into a different market.

Plant policy variables

Each firm will devise a plant policy according to a number of variables, including the following:

- size of firm
- type of work
- geographical dispersion of work.

Size of firm

This may determine whether the firm has the capital resources to acquire and use plant. It is not just the current size of the firm which is important, but also its plans for future size. If expansion is envisaged, then there may be a greater willingness to make the necessary long-term financing arrangements — assuming, of course, that financial market institutions are willing to co-operate. Indeed, increased plant usage may be a method by which a firm expands its capacity to produce, and thus enter new markets.

Type of work

A key factor is whether the type of work a firm undertakes is likely to allow sufficient utilisation of owned plant. For example, high utilisation seems more likely if:

- the firm specialises in 'plant-orientated' work such as excavation and drainage, which requires the same type of plant repeatedly
- the firm tends to undertake a fairly narrow range of work in that it always operates as a main contractor or subcontractor
- the contracts which a firm undertakes all tend to be of a similar size and type, again encouraging consistent plant usage.

Geographical dispersion of work

Plant needs to be stored, maintained and transported between sites. Whether this can be achieved economically depends on such factors as the distance between centres and concentrations of work. Therefore, if a firm undertakes most of its work in a well-defined locality, then the opportunities to service this work economically from a single plant depot are enhanced. If sites are isolated and dispersed, transport costs will be much higher.

Plant implications of expansion

When a firm changes the size and structure of its activities, plant needs may also change. There are several ways by which expansion may be achieved:

- increased number of contracts

- increased size of contracts
- more geographical locations
- diversification into other kinds of work.

Of course, these are not mutually exclusive, as the firm may well expand in different ways simultaneously. From the point of view of plant policy, the firm must ascertain the effect that different kinds of expansion will have on plant requirements. The first three methods may increase plant usage, but not necessarily require different types of plant to be used. More contracts and more geographical locations may mean more items of the same plant. Bigger contracts may mean larger sizes of plant, but if the firm is still specialising in excavation and drainage, then this defines the type of plant required. Plant for hoisting will still not be necessary. The final category of expansion, diversification, may require different types of plant to what has hitherto been necessary, and the firm may wish to think more carefully about the implications before embarking on that route.

The corollary of expansion is rationalisation. When a firm expands there may be the additional costs of more items of plant, but there may also be the opportunity to reduce costs due to rationalising plant management procedures. For example, the new distribution of activities may lend itself to fewer but larger plant depots, yielding economies of scale in storage and maintenance, and possible savings in transport costs.

Whether plant should be bought or hired

This is a perennial dilemma which does not have a definitive answer. However, it is possible to assess the general advantages and disadvantages of these two options when formulating a plant policy.

In the case of *bought plant*, the potential general advantages are:

1. Availability of required plant is maintained, thus helping the effective planning of projects and the accurate forecasting of costs.
2. Owned plant can be used as a marketing tool since the surface of an item of plant can provide a valuable space for publicity, provided, of course, that the plant is well maintained and clean.
3. As long as the plant is fully utilised, there should be a cost saving as the hirer's profit margin is eliminated.
4. Owned plant may give an opportunity for competitive marginal cost pricing. This was explained in Chapter 11, when it was shown that the fixed cost element of owned plant can be ignored in the short term, while the entire cost of hired plant is variable and must therefore be covered in pricing.

The potential general disadvantages of *bought plant* are:

1. A fixed cost element has to be covered over time, which can be a problem if full utilisation is not achieved.

2. Bought plant represents capital tied up, which is therefore not available as working capital for financing projects.
3. The additional costs of maintenance, training, insurances and premises are required.
4. There may be a tendency to use a particular piece of plant because it is owned, even if it is not the most suitable for a given task.

In the case of *hired plant*, the advantages and disadvantages are broadly the reverse of those above, although the following additional points can be made:

1. There are advantages of capital not being tied up and fewer fixed costs, plus the ability to hire the precise plant required for a task.
2. There may be tax advantages in hiring, since hire charges are business expenses, 100 per cent deductible against revenue, immediately before taxation is applied. By contrast, bought plant is a capital expenditure, offset against revenue more slowly over a number of years.
3. Some particular disadvantages of hired plant are that there may be potential availability problems in boom periods. In addition, plant hire rates may be liable to fluctuation, making accurate cost forecasting more difficult.

The decision of whether to buy or hire plant should be considered in the light of the next section, on the function of the plant company.

Role of plant company

When a firm owns sufficient plant, it is likely that a separate plant company will be set up as an element in an M-form organisational structure. As mentioned earlier, a major decision to be made is whether the plant company should be constituted as:

- a service company to the construction divisions, or
- a profit-maximising division in its own right.

If the former, then the advantages of bought plant, such as guaranteed availability, will apply. However, it may be argued that the staff of the plant company will be better motivated if they are able to pursue objectives of their own. In this case, the plant company would conduct its business in the most profitable way, even to the extent of hiring its plant to a competitor of the related construction division. This, however, would mean that the potential advantages of plant ownership would not be available. The implication is that hiring would be the only effective option for the construction division. This is not an easy problem to solve — what is advantageous for the construction division may not be in the interests of the plant company.

Quality management

When discussing organisation structures and Figure 20.1 earlier in this chapter, it was stated that responsibility for quality assurance is likely to be exercised at regional level. But the decision may have been made at a higher level that all parts of the organisation will be required to put in place quality management systems. Quality management was first considered in Chapter 10, in the context of production management. The main elements of a quality management approach are:

1. It should permeate the whole organisation.
2. Externally, it involves close, trusting and non-confrontational relationships with customers, suppliers and subcontractors.
3. Internally, all contacts should be treated as if they are valued customers or suppliers.
4. Resources should be invested to ensure that problems are designed out before they occur, and that everything is 'right first time'.

There is often a misunderstanding of what is meant by 'quality'. It does not necessarily mean something which is good. Rather, it means conforming to a standard or requirement which has been set or which is expected. Therefore, quality is about a product or service being fit for the purpose expected by the client or customer. Quality standards may be set at any level — high, medium or low. The important thing is that the outcome matches what is expected. This relates to the general management process of control, which concerns ensuring compliance with plans. The idea of a quality management system is that procedures should be put in place, so that compliance with the standard should always occur.

The importance of setting up appropriate procedures to ensure that things are right on an ongoing basis is a concept which has other applications in management. For example, in Chapter 17 on industrial relations, it was shown how a procedural collective bargain agreed and implemented over a number of years can be a significant help in ensuring that the annual bargaining on substantive issues, such as pay, can proceed more smoothly. Thus, the procedural collective bargain can be seen as a way of implementing trusting relationships within the organisation.

As discussed, quality management is concerned with the organisation taking a good hard look at itself, and considering how it might improve what it does. There may well be benefits in terms of increased efficiency and productivity, and cost savings.

Within the construction industry there is great interest in quality assurance. Although there are overlaps, quality management is primarily about the organisation looking at itself, while quality assurance is primarily about others looking at the organisation. Sometimes this might be a client or customer, but more likely it will be a third party who will certify that an organisation has appropriate procedures in place. The third party body best known in the UK construction industry is the British Standards Institution, and their BS 5750, Quality Systems. As stated in Chapter 10, firms who have achieved quality assurance certification under this standard have shown that they have developed the necessary framework and procedures to enable a quality management system to be established.

As mentioned above, there may be cost saving benefits to be gained by quality management, but the construction industry's interest in quality assurance may be motivated by different reasons. In a study made by Pateman in 1986, it was found that manufacturers and suppliers of building products and materials did not expect any financial benefits from quality assurance certification. They did not expect to reduce costs while maintaining quality, nor to improve quality while maintaining costs. What they did expect to gain was some sort of marketing advantage over their competitors, including those from overseas. Pateman appeared critical of this idea, claiming that marketing advantages only occur from competitive pricing. He argued that the real benefits of quality assurance derive from the rule book which documents the system required to control the quality of a product or service, and to which all employees can subscribe.

Many would probably dispute Pateman's downgrading of the potential marketing advantage of quality assurance. Construction firms in both the private and public sectors have gone down the path of quality assurance certification, and the marketing aspect of satisfying customer and client needs is often emphasised — see, for example, Morrison (1995) writing on behalf of a private firm, and Preston (1995) on behalf of a direct labour organisation. In the case of Morrison, there is particular emphasis on the possibility that quality management can help create the environment in which partnering will flourish.

A quality system which meets BS 5750 requires a good deal of work to set up. There can be much documentation, which sometimes leads to the criticism that quality assurance is very paper-orientated and merely requires numerous ticking of boxes in order to comply with its requirements, while nothing important really changes. However, supporters argue that systems have to be monitored, maintained and improved to retain certification. Firms with quality management systems in place have a quality manual which details procedures at all stages of its operations. This could include design, procurement and construction. Therefore, it would appear that construction firms offering a more complete service to clients, perhaps including partnering arrangements, would have most to gain from quality management systems. This does tend to emphasise the purpose of quality management as being to meet customer needs and therefore is consistent with the market-orientated firm discussed at various times in this book.

Health and safety

It is well known that the safety record of the construction industry is very poor. Some may say that the industry is inherently dangerous because of the nature of the technology, exposure to weather conditions, temporary structures and the like. However, it is also possible to argue that there is no such thing as an accident, because an incident only occurs if the firm's health and safety policy is not stringent enough, or because the correct safety systems were not in place, or because of some human error, which by definition is avoidable.

In assessing why there are so many deaths and injuries in construction, the methods of site management and organisation should not be ignored. For example, although the contractor may have a safety policy and employ safety officers, the majority of work may be subcontracted out. Therefore, the operatives are not under the direct control of the main contractor. If there is also a payments-by-results system in place, which is quite common, then strict adherence to time-consuming safe procedures may lead to a reduction in income. In addition, there is often the so-called 'macho' tendency, where peer pressure might mean that individuals do not take appropriate personal safety precautions for fear of losing face with others. The more competitive the market situation, the greater the tendency to cut tender prices and cut corners in general. This could have implications for health and safety unless there is a strictly enforced legislative framework which applies universally, and prevents anybody obtaining a competitive price advantage by cutting corners on safety.

Although there is a tendency to bracket health and safety together, with poor safety receiving most attention, they nevertheless deserve separate consideration.

Health

Health hazards have not received the same attention afforded to safety. This is partly because the health risks attached to the use of certain materials and methods may not have been fully realised at first. Fryer (1990) identifies a number of physical health hazards:

- *Dusts* These include silica and asbestos dust, as well as cement dust. Even where substances are no longer used due to the health aspect, as is the case with blue asbestos, they may still be encountered in demolition and renovation work.
- *Toxic fumes* These can be a problem in conjunction with painting, welding, flame-cutting and lead-burning. As with dusts, the extent to which the workplace is confined will be a factor.
- *Vibration and noise* The use of heavy, noisy, vibratory tools and equipment can lead to a range of health problems including numbness, hearing damage and back problems.
- *Skin troubles* These arise from a variety of materials such as cement, tar, solvents and acids. Individual reactions to these substances vary.

It should be emphasised that exposure to many of these hazards can be reduced if the firm has a policy to ensure that appropriate protective clothing is available, its use is compulsory, and that maximum exposure times are scrupulously followed.

Safety

There are a number of ways in which incidents leading to death or injury can occur, including:

- people falling
- collapse of structures and excavations
- objects falling on people on or off-site
- machinery collapses such as hoists or cranes
- vehicular accidents in connection with wheeled plant.

It is recognised that some trades are at greater risk of injury than others. These include steel erectors, demolition workers and scaffolders. However, as in the case of health, many of the problems can be avoided if it is the firm's policy to treat safety as a priority, even if it does result in additional cost. In this respect the influence of clients and their consultants could be vital. For example, they could insist on examining the health and safety record and policy of any contractor they intend to engage, and be prepared to pay an appropriate price premium. If this was allied to effective legal regulation which ensured that all contractors implemented, and incurred the cost of, effective health and safety policies, then many of the problems would be eliminated.

Legal regulation of health and safety

Although this exists, it does not achieve the results hoped for in the previous paragraph. This might be because:

- the laws themselves are inadequate
- the laws are not enforced rigorously enough.

A brief review of the legal framework will follow. The law can be considered under the following headings:

- Common law
- Health and Safety at Work etc. Act
- Construction Regulations
- Construction (Design and Management) Regulations.

Common law

The law of tort applies, especially negligence. Thus an employer owes a duty of care to the workers on site. There is also a duty owed to those off-site. These include members of the public likely to be affected by the works. The legal duty of the employer may apply through the doctrine of vicarious liability. The existence of this law does not absolve the worker from responsibility, since the employer has to take 'reasonable' care. Furthermore, if an accident is adjudged to be partly the fault of the worker then contributory negligence comes into play.

Health and Safety at Work etc. Act 1974

The common law provides that compensation can be paid if an accident occurs, but does nothing positive in itself to prevent accidents and create a safer working environment. This major piece of legislation seeks to remedy this. It derives from the work of the committee chaired by Lord Robens in the early 1970s. There are a number of features and provisions of the Act, including the following:

1. It aims to make employers take positive action to improve safety.
2. It provides for criminal as well as civil liability.
3. It covers virtually all employees.
4. It imposes specific duties on employers and employees.
5. It requires every firm to prepare a safety policy.
6. It set up a machinery for implementing the Act, particularly the Health and Safety Commission, and Executive.
7. It gave considerable powers to health and safety inspectors, including, in the last resort, the power to close a workplace.

Many of the criticisms of the Act have not necessarily been about the provisions themselves, but about the inadequate numbers of health and safety inspectors employed to enforce them.

Construction Regulations

The Act described above lays down general duties, while these regulations give detailed provisions for use on construction sites. There are four sets of regulations dating from the 1960s, as follows:

- **General Provisions** These include regulations regarding the appointment of safety supervisors, and matters such as safety precautions for excavations and the like.
- **Lifting Operations** These cover the maintenance and use of lifting devices such as cranes and hoists.
- **Health and Welfare** These lay down minimum standards for shelters, together with eating, washing and toilet facilities on site.
- **Working Places** There is special emphasis on places such as scaffolding, ladders and walkways.

Other regulations may also affect construction work.

Construction (Design and Management) Regulations 1994

The CDM regulations, as they are known, actually came into force on 31 March 1995. They place new statutory regulations, backed by criminal sanctions for non-compliance, on parties who may not have been specifically mentioned in previous legislation. There are five key parties, who may be firms or individuals, who have specific duties. These are:

- the client
- the designer
- the planning supervisor
- the principal contractor
- contractors and the self-employed.

One of the interesting features of this is that it recognises the role that clients could have in making sure that their projects conform to good health and safety practice. The role of planning supervisor is new, and involves responsibility for co-ordinating the health and safety aspects of the design and planning phase. This role is also responsible for the pre-tender health and safety plan, thus requiring these aspects to be considered at design stage. The principal contractor is obliged to take over and develop the health and safety plan and co-ordinate all contractors so that they comply with the law on health and safety. Along with the other duties, this should encourage construction firms to think carefully about their health and safety policies. The CDM regulations seem to require a significant effort from all team members, and time will tell whether this will lead to an improvement in the construction industry's health and safety record.

Summary

This wide-ranging chapter has considered various aspects of construction firm policy. Given that most firms aim to be market orientated, the organisational structure chosen will tend to be multi-divisional or M-form. From this, three policy areas were selected for discussion: plant management, quality management and health and safety policy. Although these were mainly discussed from the point of view of managing the firm, they ultimately affect the management of construction projects, which is the subject of the next chapter.

Further reading

This chapter has referred to previous chapters and hence to the further reading contained therein. Apart from references made in the text of this chapter, additional information is as follows.

For plant management, see Harris and McCaffer (1991), which is a paper in the CIOB Technical Information Service, and Smit (1990). For quality management (in addition to those cited in the text), see Monaghan (1987) and Lai (1989), both papers in the CIOB Technical Information Service. For health and safety (in addition to those cited in the text), see Fryer (1990), ch. 13, and publications on CDM from the Health and Safety Executive.

21 Management of construction projects

Introduction

In this and the subsequent two chapters, the emphasis will move from the management of firms towards the management of projects. In this chapter some of the processes of management examined in previous chapters, such as Chapter 8, will be more specifically applied to the construction context. One of the distinguishing features of projects, as opposed to general management, is that they occur within a given time span. Therefore, the main focus of this chapter will be the management of time, before moving on to the management of costs and production in subsequent chapters.

Having reviewed the elements of project management, attention will then turn to the needs of the client — the initiator of the project. Clients undertake projects because they satisfy some wider organisational need. This must be understood by the project team who must deliver the project within prescribed cost, time and performance requirements. As mentioned above, a feature of projects is that they take place within a given time span. Therefore consideration must be given to the planning of projects — who does it, how it is done, and how the resulting information will be used. The focus will be on time, but with some consideration given to trade-offs between time and cost. Finally, some of the sources of project management data will be examined.

Elements of project management

The concept of project management, as distinct from the general processes of management, was introduced in Chapter 8. The main points can be summarised as follows:

1. Projects take place in a given time span, and therefore need to be carefully planned with realistic intermediate targets or milestones.
2. A project team, likely to be multi-disciplinary, will need to be set up for the purpose.
3. Many construction projects represent a considerable capital investment carrying a high degree of risk.

4. The allocation of this risk between client and contractors can lead to complex and fraught contract arrangements.
5. For the contractor, and for some professional clients, each project may form a significant proportion of total workload, thus accentuating risk and causing a potential problem of peak loading of resources.
6. The project team needs to be fully aware of the wider organisational needs of the client and of any subsequent user or occupier.

To return to the time element, because each project has a specific time span, it follows that the project will proceed through a number of stages. These stages are not discrete and may take a slightly different order. Furthermore, there is no definitive project framework, as each participant or commentator will have his or her own view, which may vary slightly. As an example, the six stages of a project, as identified in Chapter 8, are:

- the idea for the project
- the decision to proceed
- design
- procurement
- construction
- occupation and use.

Before considering the planning and control of project time in more detail, it is appropriate to pause and reinforce the predominance of client needs in the management of construction projects.

Client needs

Satisfying customer or client needs has been a theme of this text and is a vital requirement for firms. For example, in considering the market strategies of construction firms in Chapter 19, emphasis was placed on the quantity and quality of demand, as well as methods of delivering demand through the various procurement methods. The objectives of clients were first discussed in Chapter 3, which put the objectives of the project into the broader context of the client's organisation. To relate this to the framework above, clients may have an idea for a project based on a perceived market opportunity, or on the duty to satisfy a social need. For example:

1. The commercial client, building to sell or let space to others, is expecting a direct financial gain.
2. The manufacturer is expecting a productivity gain from building a new factory.
3. The public authority is expecting a social investment gain from a new school or transportation initiative.

When the client is satisfied that there is a need for a project, the next stage is to undertake the necessary feasibility studies to ascertain whether the proposed project will meet the objectives of the client. Feasibility studies will include technical and financial elements, and, as a result, an investment decision will be made whether to proceed. If so, project objectives will be specified in terms of time, cost and quality along the lines discussed in Chapter 8.

With the decision to proceed taken, project planning within the time, cost, and quality parameters can commence in earnest. In particular, work can begin on:

- developing the design, which aims to produce a workable solution to satisfying project objectives, and
- deciding on the most appropriate procurement methods, in terms of contractual arrangements and the selection of consultants and contractors

It should be noted that the project management expertise employed up to this point may have been within the client organisation, or through external consultants.

Project planning and control

As stated above, once the decision to proceed has been taken, planning the project can begin in earnest. This topic is usually associated with time management, but the same information can also be used for considering cost. When referring to the planning of time, the term 'programming' is often used. The general rules for control, as a process of management, apply; that is:

- targets are set
- actual progress is monitored
- remedial action is taken if necessary.

To gain an understanding of what is involved with project planning and control, the following matters need to be considered:

1. Is there a need for planning and programming?
2. What processes are required to draw up a plan or programme?
3. To what use can a programme or plan be put?

Each of these will be examined before considering some of the techniques used for planning and programming.

Need for planning

Once the decision to proceed has been taken, the client or client's representatives will need to set up an ongoing planning procedure. From the time contractors become involved, say, from being invited to tender, they will need to draw up a plan, if only to help prepare a more accurate estimate. The contractor who is finally awarded the contract will need to plan in detail for a variety of purposes. In the light of the framework for a project previously described, planning at the following stages will be considered:

- design stage
- procurement stage
- construction stage.

Design stage

This, of course, can be quite a lengthy process, initially produced in outline form before eventually becoming detailed enough for tendering purposes. The very early stage of the design process, just after the decision to proceed, is a critical time for the development of the client's brief. Hamilton (1990c) identifies this particular period as the project concept phase, and defines certain tasks to be carried out as part of the Project Master Plan (PMP). These tasks are seen as essential elements of project management, and include:

- developing the project's scope of works
- managing the work of designers in the production of the conceptual design
- producing the project's master time schedule
- developing the overall budget estimate
- establishing the appropriate contract strategy for the project.

Hamilton believes that this concept stage should never be hurried, as getting these major strategic decisions right at the beginning will pay dividends throughout the project. In examining the above tasks, it can be seen that matters of time, cost and quality are dealt with, as well as laying the foundations for successful detail design and procurement. In respect of time, the master time schedule needs to lay down important milestones from the client's point of view. This can be achieved by working backwards from final completion. This may be defined as when the building is delivered, and can begin to make a contribution to the objectives of the client organisation.

Procurement stage

At this stage, both client's and contractor's organisations will need to plan. The client will review the elements of the PMP discussed above to ensure, in the light of

- detail design
- updated cost estimates
- negotiations with and/or estimates received from contractors

that time, cost and quality targets are being maintained. The contractor will need to collect all relevant information to prepare an accurate estimate. This will obviously include cost data in respect of labour, materials, plant and subcontractors. A time schedule or programme will also be required to ascertain the indirect costs of running the site, lead-in times for the delivery of components, and how the various work packages will fit together.

Construction stage

The client will continue to ensure that the various targets are being met. In particular, it is important to avoid cost and time over-runs. The contractor will engage in detailed planning. There is normally a requirement under the contract for a project programme to be supplied to the client. The contractor will also generate a good deal of information for internal use. Plans are likely to be prepared for a number of time-scales. For example, the contractor will need to know broadly what materials and components have to be ordered, and which subcontractors need to be engaged over a period of months. In a more detailed vein, there will also need to be plans for the current week, in respect of matters such as the use of site labour, and the deployment of major items of site plant such as the tower crane. The contractor will need to relate progress to cash flow to ensure that financial targets are being met.

Some of the points raised in this section will be further considered in the next two chapters on financial and production management.

Process of planning

According to Hamilton (1990d), a programme is developed in three steps:

- modelling
- estimating
- balancing.

Modelling

This is very much a qualitative exercise, involving the definition of all the items of work and activities covered by the programme period, and ascertaining how they relate to each other. This will include noting which activities must precede others. It is important that all activities are included. At this stage, there are no time or resource constraints to worry about.

Estimating

The above model now needs to be quantified. Information is added in respect of time taken for each activity. Also required is an analysis of what resources are needed, and when they will be required. Finally, it should be possible to derive an idea of what the costs are likely to be.

Balancing

Adding the estimate to the model is quite likely to result in an unbalanced picture, with perhaps bottlenecks in the supply of resources, and excessive amounts of activity and resource deployment at certain times. Balancing enables a better use of resources and the achievement of appropriate trade-offs between cost and time.

Use of plans Before considering the types of planning techniques available, it is worth identifying some of the information which may be extractable from plans and programmes. The list could include:

- duration of each activity at different resource levels
- total duration of whole project
- sequence of activities
- when resources will be required
- which activities are critical, defined as those which, if delayed, will delay the whole project
- what trade-offs may be available between time and cost.

It should be noted that not all of this information can be obtained from all types of programme. There may be a trade-off between ease of use and the amount of information obtainable.

Planning techniques

There are a number of planning techniques in use, but a certain number of problems exist in making them work effectively.

Work on construction sites is notoriously difficult to model accurately because of potential interruptions due to weather conditions, diverse trades and subcontractors, and the often haphazard storage, movement and use of materials. This problem has reduced in recent years with more mechanisation and use of manufactured components. Nevertheless, work on a construction site can never be as easy to plan as in an enclosed factory.

Contractors have not always been effective at collecting reliable data and disseminating it as appropriate throughout the organisation. Different contracting functions, such as estimating and planning, often collect data independently for their own particular purposes. Sometimes 'the planners' are regarded as remote from site management. What they produce is usually available in the site office, but may not be rigorously adhered to. It is the task of senior management to ensure that internal rivalries do not detract from the effectiveness of the organisation.

It is important that planning techniques should be an effective instrument of control. This means that the technique should enable:

- the setting of targets
- the monitoring of progress
- information to be capable of being acted on quickly, in order to remedy any problems revealed.

Thus targets should be realistic and flexible. The more widespread use of computer programs has enabled plans to be more quickly updated.

The two types of planning techniques in most common use are:

- bar charts
- networks.

Bar charts

These are also known as Gantt charts, named after Henry Gantt, who is credited with their invention. Gantt was a disciple of F.W. Taylor, and therefore one of the early exponents of scientific management, particularly in respect of time management. Bar charts can be seen on almost every construction site. They represent the most straightforward and widely used method of depicting a project programme. Examples of bar charts are shown in Figures 21.1 and 21.2. Figure 21.1 is a programme drawn up by the project manager at early design stage, or concept stage, which shows the client the main blocks of time leading to completion. Figure 21.2 is a programme drawn up by a contractor to show the main activities in constructing a small building.

The main features of a bar chart are as follows:

1. The activities are listed on the vertical axis.
2. Time is shown on the horizontal axis. This may be any appropriate time unit, such as months or days.
3. Each bar is divided horizontally into two, with the top half depicting when the activity is planned to take place, and the bottom half shaded in, as proportions of the activity are completed.
4. A vertical movable marker line indicates the current time, thus enabling the observer to note if any activities are behind schedule, and to what extent.

If observation of the bar chart reveals that actual progress does not match planned progress, then the alternative courses of action are:

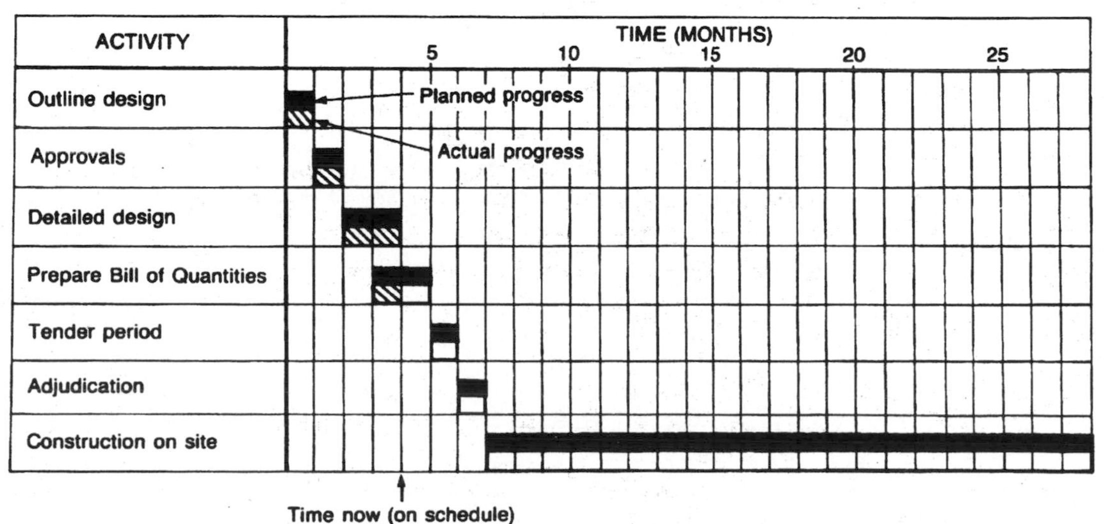

Fig. 21.1 Programme drawn up at concept stage

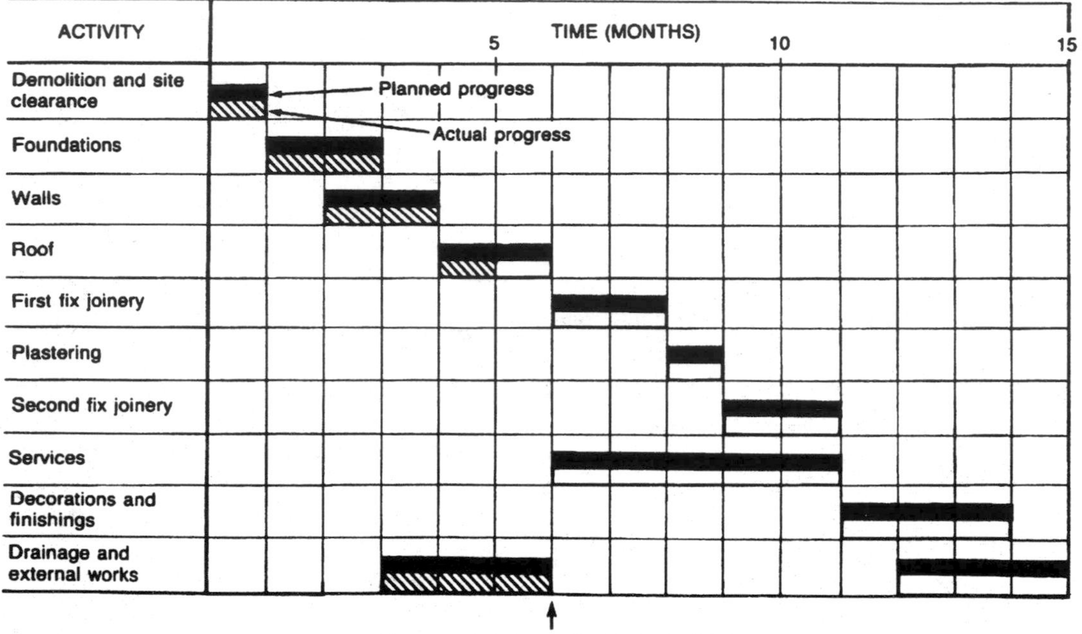

Fig. 21.2 Programme for construction of small building

1. The programme should be adjusted to fit the existing time schedule.
2. The schedule should be extended because it is now felt that some of the original activity durations were unrealistic.

Some of the reasons why time may need to be saved, together with methods for achieving same, will be examined later in the chapter.

Networks

These are more ambitious than bar charts in the information they seek to present. They are also less straightforward to understand and are consequently referred to less in practice. Nevertheless, the availability of computer software has increased the use of networks, particularly since they can be updated more easily, and because bar charts can be prepared from them.

An example of a network is shown in Figure 21.3. The main features of networks are:

1. They have the advantage of showing more closely the relationship between activities, in terms of what must be finished before something else commences.
2. The lettered circles represent the beginning and end of activities. Therefore activities are specified by the beginning and finishing circle, for example, A–B, B–E.

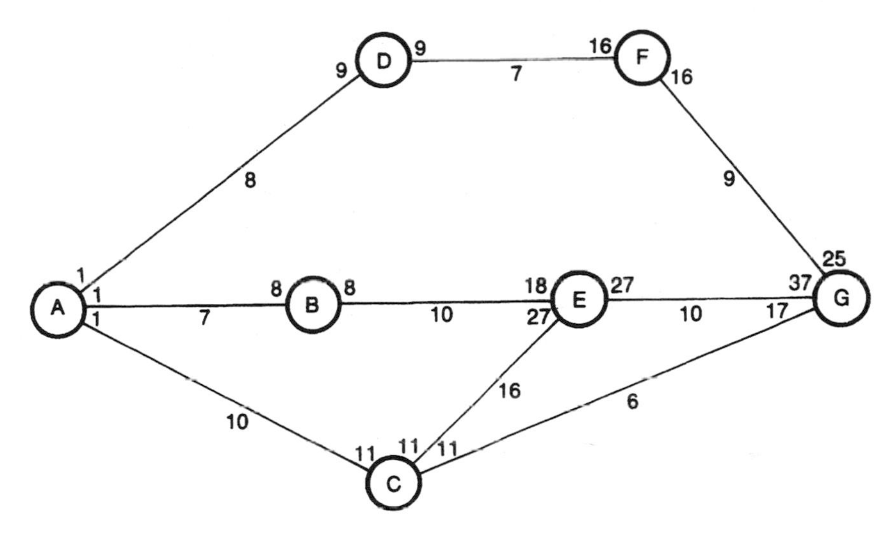

Fig. 21.3 An example of a network programme

3. The figure shown on leaving a circle signifies the earliest time an activity can start, while the figure shown on entering a circle shows the earliest time an activity can finish. The figure on the line between the two circles represents the duration. Therefore, the duration of activity A–B is 7, with its earliest start being 1, and its earliest finish being 8.
4. The critical path can be highlighted. This is the longest path through the network, in this case, A–C–E–G. It is critical because a delay to any activity on this path will delay overall completion time for the project.
5. Whereas activities on the critical path must be carried out at particular times for the programme to remain on schedule, there is some choice as to when non-critical activities can be carried out. For example, A–B and B–E together take 17 days, but can be carried out any time within days 1–27. This period of choice is known as the 'float'.
6. Assuming that appropriate data is available, networks can be used to explore time/cost trade-offs, and can test different levels of resources. Project management data will be examined towards the end of this chapter.

Figure 21.3 shows a fairly straightforward network which illustrates the principles. They can be much more complicated than this, for example, to allow for probability variables. The idea of networks was developed to help plan major military projects. In construction projects many people feel that bar charts retain the virtue of simplicity and are therefore to be preferred.

Time management

It is one thing to apply planning techniques in a mechanical way, using computers and given data, to derive a programme for a project. It is another matter entirely to understand the importance of time, and what management decisions have to be taken in order to implement a required change to the schedule. Therefore, the matters to be considered in this section are:

- the circumstances in which a decision might be made to speed up the programme, and achieve faster completion than that currently expected
- the general management approaches which are available to achieve faster completion.

When faster completion is required

The decision to speed up a project may be made by the client or contractor.

Client

It has already been established that clients require projects to be completed in accordance with specified time, cost and quality objectives. To some clients, if the priority is to keep costs down, then a delay may not matter too much. However, to other clients, time may be an absolute priority — for example, when a shopping centre is scheduled for completion to meet a seasonal increase in trade. The usual contract conditions specify that if there is a genuine delay which is the fault of neither party, then the contractor is entitled to an extension of time. In these circumstances, if the client wishes to adhere to the original completion date, then the client would have to authorise variations to the contract, which may have cost implications.

Contractor

If a delay occurs for a reason which does not allow an extension of time under the contract, then the contractor will be liable for liquidated damages. If these are set at a high level — which is quite likely in cases where a client treats time as a priority — then delay will be expensive, and the contractor will probably decide to speed up operations in order to meet the original completion date. In general, the contractor's decision on whether or not to accelerate the programme will depend on the comparison between any additional production costs on the one hand, and the extent of liquidated damages on the other.

Approaches for achieving faster completion

There are a number of ways in which management may seek time savings, with varying implications for resource usage and costs. They may be classified as:

- improving site performance
- employing additional resources
- changing method of production
- changing design.

Improving site performance

This requires no additional resources. If it is felt that the existing resources are not performing as well as they might, there may be scope for improvement. Changing site management may help, for example, to improve industrial relations on site. Perhaps the control of subcontractors, or the flow of materials, is not as good as it should be. Improved financial incentives might be offered to operatives. A better understanding of what motivates people or improves teamwork, as discussed in Chapters 15 and 16, might also be of assistance. Whether sufficient improvements could be achieved with these measures part-way through a project depends on the scale of the problem and the amount of time which needs to be saved.

Employing additional resources

This refers to a quantitative increase in resources. In other words, more of the same production methods and resource types are being used — additional machines; additional gangs of workers; longer working hours. Many textbook examples which show time and cost data for different resource levels tend to assume similar types of resource. It may sometimes be possible to accelerate the programme sufficiently by employing additional resources, but at other times this may not be practicable, for example:

- the location of the site may not be easily accessible
- the nature of the problem activities may make it physically impossible to deploy additional resources (as would be the case with tunnelling)
- market conditions may determine that additional resources are simply not available except perhaps at a prohibitive cost.

Changing method of production

This is a more radical solution which may involve changes to quantity and quality of resources. Examples include using more and bigger plant, using faster production techniques, such as pumping concrete to higher levels of a building rather than lifting by hoist. Changing methods in this way will not necessarily be easy during production, although construction is more flexible than manufacturing in this respect. It is best to incorporate more efficient production methods at an earlier stage of planning.

Changing design

This is the most radical solution, and therefore often the most difficult to implement once a project is under way. It could involve changing the specification of materials and components to introduce more prefabrication. There may be some scope for this, although much may depend on the availability of the components. There is little point in saving site time through prefabrication if the time taken to receive the components is correspondingly increased. It is very unlikely that significant changes can be made to structural elements, but changes may be possible in respect of finishings and fitting out. One example is for prefabricated toilet or washroom 'pods' to replace individual appliances in buildings such as office blocks or student halls of residence.

In summary
Assuming that the fastest method was not planned at the outset, then by using one or more of the above approaches, the programme can be compressed to achieve an earlier completion time.

Time/cost trade-offs

Costs will be considered in more detail in the next two chapters, but here it is appropriate to consider certain aspects of the relationship between cost and time. It is usually assumed that building more quickly costs more. Therefore, the client who wants to open a new shopping centre in time for Christmas will expect a higher price tag. In the cases described above, where the client or contractor decides to accelerate the programme in order to finish on time, it is very likely that costs will rise. However, in the scenario where a fast completion is planned at the outset, the situation is not so clear cut. This is due to the structure and classification of costs.

In Chapter 11, costs were classified as:

- variable, which are dependent on levels of production
- fixed, which are incurred irrespective of levels of production
- a whole grey area of semi-variable costs in between.

A similar classification could be made for construction project costs. These are often summarised in a bill of quantities, where:

- variable costs are described by the measured items — square metres of concrete bed, linear metres of pipework, etc.
- fixed costs are normally included in preliminaries — setting up and clearing site, connection to services, etc.
- semi-variable costs are also normally included in preliminaries, and often consist of items related to time — hire of site huts, supervision, etc.

If the client's choice is for fast completion, then the approaches for achieving this will be explored — more mechanised construction methods, more prefabrication and so on. This is quite likely to prove more costly in terms of variable, or direct costs, as they are sometimes called. However, there will be a saving in time-related preliminaries, which may go some way towards off-setting the higher variable costs. Thus, a 'menu of choice' in planning emerges, as depicted by Figure 21.4. It can be seen that the fixed and time-related preliminaries curves commence from time zero. However, the variable costs or measured items curve will commence at time t_1. At this point, maximum advantage has been taken of time-saving methods of completion described above. The total cost curve is a summation of the other three graphs, and therefore can be regarded as a planning tool to ascertain the trade-off between time and cost.

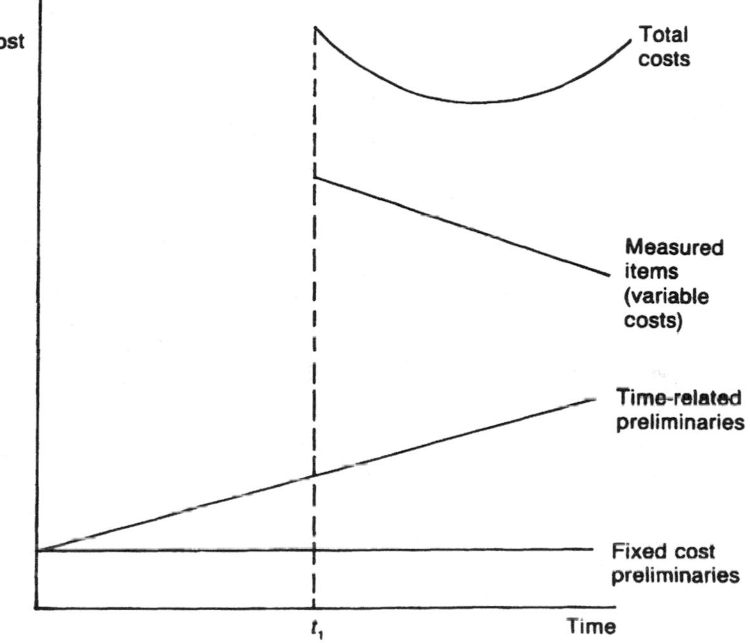

Fig. 21.4 Time/cost trade-offs: a 'menu of choice'

Project management data

The above described planning problem of relating time to cost, or indeed any project planning problem, requires the availability of appropriate data if practicable solutions are to be found. Techniques, no matter how theoretically sound, can only be applied if there is adequate data. There is, of course, a wealth of cost data on things like materials prices, plant hire charges, etc., and quotations from subcontractors can be obtained fairly easily. Data on standard times for carrying out particular activities, such as digging a cubic metre of soil, also exists. The problem with this sort of data is that, for any given activity, there may be numerous ways of carrying it out, using different combinations of plant and labour. Indeed, as has been shown above, the speed of construction can vary depending on at least four approaches by management. This includes the case where the same quantity and quality of resources can perform in very different ways, depending on the performance of management, and the motivation of the workers.

It was mentioned earlier that network methods of planning allow different resource levels to be tested in order to ascertain their effect on the programme, particularly on project duration. Examples sometimes shown in textbooks include data on time and cost for 'normal' resource levels, and 'crash' resource levels. The latter is defined as flooding the project with resources, irrespective of cost, to complete the activity in the fastest time possible. The problem with this approach is that it assumes that the only way to accelerate the programme is to employ more of the same type of resource. However, as established earlier, greater speed can also be achieved by utilising the existing resources more effectively, changing the method of production, or even changing the design. Data on all these possible options may be difficult to obtain and keep up-to-date, but it may be necessary to make some attempt, in order to plan accurately.

Another problem with the data and numerical examples sometimes shown in books is that they tend to assume traditional methods of project organisation, where the main contractor directly employs and controls a labour force. However, in reality the project is more likely to be divided into a number of work packages, most of which are subcontracted out. In addition, greater use is often made of prefabricated components. This means that an increasing amount of the 'value added' in buildings is created off-site. Therefore, the issue for planning is no longer to predict how many units of labour will be required. Instead, planning now concerns the process and time-scale for the procurement of subcontractors, and the ordering of materials and components. Published data exists on these matters, such as the monthly 'Foresite' series in the journal *Building*. This data shows lead times for the delivery of 38 key building materials packages such as steel frames, brickwork, cladding panels, metal windows, passenger lifts, air conditioning plant, and so on. This is essential information when planning a project. The lead time between order and delivery for each material depends on economic conditions. This includes what is happening to demand for each material, and the extent to which supply can readily adjust to it. In the boom of the late 1980s, demand rose significantly while the inelastic supply of certain components, such as cladding panels and passenger lifts, resulted in long lead times (as well as major price increases). In some cases, lead times exceeded a year. In a recession, lead times tend to shorten, but this depends on how well supply adjusts to demand. There is always the danger that if supply capacity is reduced, bottlenecks in supply (and rising tender prices) will occur as soon as demand rises.

The same is true for the supply of capital goods in the economy generally. Capital goods are normally inelastic in supply, which means that not everybody can acquire them at the same time.

Summary

This chapter has been concerned with project management. Having started with the basic elements, it is important to understand that the needs of the client are pre-eminent. Successful project management depends on effective project planning and control. The need for, and the process of, planning was considered, before assessing some of the planning techniques in common use. The core of planning is time management. The reasons for speeding up a project were considered, together with the options for achieving faster completion. Although the main focus was on time, the relationship between time and cost was also considered. Finally, the sources of data available for effective project management were examined.

Further reading

Articles on project management and planning often appear in journals. See, for example, Hamilton (1990a–f), Cole, J. (1990), Mace (1990), France (1993) and the 'Foresite' series in *Building*. Relevant Construction Papers of the CIOB include: Baxendale (1992) and Harrison (1993). Various construction management books deal with planning in some detail, giving numerical examples. See, for example, Cooke (1992) and Calvert *et al.* (1995), particularly ch. 16.

22 Financial control of construction projects

Introduction

The previous chapter commenced the examination of the management of construction projects, with particular emphasis on planning, programming and the control of time. There was some consideration of cost, particularly in respect of the trade-off with time, but the main emphasis on cost will commence in this chapter. The terms 'financial control' and 'cost control' may be used in a variety of ways. For this reason, the first part of this chapter will define their meaning in the context of the remaining chapters of the book. One of the themes of this book has been that most organisations in the construction process are profit-seeking firms, who also have to maintain liquidity in order to survive. Therefore, this chapter will pay particular attention to those aspects of financial management which affect the ability of firms to generate profitability and liquidity — the payments chain, scope for alterations to contract sums, liquidity problems and solutions for various parties to the project, such as the client and the contractor.

The meaning of financial control

The financial objectives of organisations are, of course:

- to make a profit to ensure long-term survival and growth, and
- to generate sufficient liquidity and cash flow to ensure short-term survival.

These objectives were first defined in Chapter 3. When discussing the nature of profit it was shown that to achieve this, the firm must successfully complete, and consistently renew, the three-stage cycle of production. This consists of:

- buying resources
- producing goods or services
- selling those goods or services.

This was illustrated in Figure 3.1. It was also shown that profits are an arithmetical derivation, arrived at by deducting the costs incurred in bringing goods and services to market from the revenue raised from selling them. In other words, profit equals revenue minus costs ($P = R - C$).

Within the three-stage process, the revenue generated depends on:

- for the housebuilder, price of houses multiplied by quantity sold
- for the commercial office developer, rents charged, say per square metre
- for the contractor, payments received from the client.

These all derive from market transactions and the flow of funds. The incurring of costs depends on:

- the price paid for resources such as labour, materials, plant and work subcontracted out, and
- how effectively these resources are used in the production process.

The first type of costs, the price of resources, are also based on market transactions, while the second type are derived from production management and productivity performance.

This chapter is concerned with financial matters which derive from market transactions. The next chapter concerns matters deriving from the production process itself.

In the context of the financial control of construction projects, market transactions play an increasingly important role for the contractor. Traditionally, the contractor would control production closely by directly employing a labour force. This gave site management a great responsibility to control costs. However, the contemporary approach to contract management is to divide the work into work packages, each of which is normally subcontracted out. Thus, in order to control costs, the contractor relies rather more on competitive strength when buying in materials, components and subcontracted services than on the efficiency of production. Production efficiency is increasingly the responsibility of subcontractors, although the main contractor's site management still have an important co-ordinating role.

These trends have had an important impact on the roles played in the construction process. The contractor's quantity surveyor has always been concerned with managing payments, both from clients and to subcontractors. As the amount of work subcontracted out has increased, the position of the contractor's quantity surveyor has been enhanced. This is because profit and cash flow is now more dependent on market transactions regulated through commercial contracts.

The payments chain

As described above, financial control of projects tends to focus on market transactions, that is, on matters such as:

- the nature of the initial transaction, in terms of agreed prices, etc.

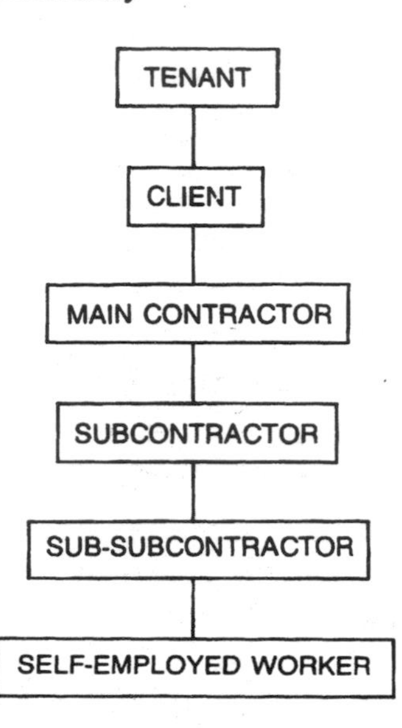

Fig. 22.1 The payments chain

- how the transaction is managed through to completion, in terms of stage payments, changes to the original agreement, etc.

The importance of contract law is apparent, because market transactions are economic bargains which are regulated by a contract. In the case of construction projects, these transactions can be complex. Underlying this series of transactions is the payments chain.

The payments chain was first introduced in Chapter 3, when it was identified as a key factor in determining liquidity — firms may be vulnerable to cash flow difficulties if they occupy an unfavourable position in the payments chain. The payments chain can represent a lengthy and complicated sequence of commercial or market relationships, as shown in Figure 22.1.

The core of the chain is, of course, the relationship between client and main contractor. As previously mentioned, the main contractor tends to fulfil a managerial and co-ordinating role, and most of the work is divided into work packages and subcontracted out. On large projects, many of these work packages can be quite substantial in their own right. They are often further divided and let to sub-subcontractors and so on. The operatives actually carrying out the work may be employed by these subcontractors, or more likely they will be self-employed (which to all intents and purposes means a one-person subcontractor, or 'independent contractor' in legal terminology). At the other end of the chain, the client often builds for others to occupy. Therefore, the chain may well be extended in this direction to include the commercial relationship between client and tenant.

Having identified the extent of the payments chain, it is necessary to assess the impact it can have. In financial terms, the relationship which exists at each link of the chain — between client and contractor, between contractor and subcontractors, etc. — affect:

- *Profitability*, which mainly depends on the price agreed between parties linked in the chain, and
- *Liquidity*, which mainly depends on the terms and conditions for payment — that is, the time difference between paying creditors and receiving payment from debtors.

Given that a firm's position in the payments chain affects its profitability and liquidity, then the factors that determine this position must be considered. Undoubtedly, the most influential factor is *market conditions*, as measured by the degree of monopoly, discussed in Chapter 4. It is also the case that the market structure of the construction industry has changed, as discussed in Chapter 19. One of the changes has been that main contractors have developed considerable market power compared with the many subcontractors competing for work. Only in times of exceptional shortages of resources do subcontractors acquire significant market power. The relative competitive strength between client and contractors is less clear cut. Some clients are large enough to possess market power, but generally speaking competitive relations are more variable, depending on the state of the market.

Comparative market conditions also affect contract conditions. A construction contract not only specifies price, but a whole host of other matters which determine profitability and liquidity. These include:

- rights to interim payments
- entitlements to additional payments and extensions of time.

The industry normally uses standardised contracts. However, if one party has greater market strength, that party may insist that clauses be added, deleted or amended. For example, during the recession of the mid-1990s, some clients were able to delete the inclement weather clause. This meant that the contractor was unable to obtain an extension of time for inclement weather, and had to take on this additional risk.

Another example which occurs in some contracts between main and subcontractors is the controversial 'pay when paid' clause. This absolves the contractor from paying money owed to subcontractors until payment is received from the client. This practice of making the discharge of obligations under one contract dependent on the performance of another contract has been widely criticised. It is unlikely that any subcontractor would accept such a clause unless forced to do so because of a significant competitive disadvantage.

Alterations to contract sum

At feasibility stage, the client allocates a certain sum of money for construction costs. The client expects to sign a contract based on that sum. Once an agreement has been reached with a contractor, one of the main tasks of financial control is to ensure that the contract sum is adhered to.

However, there are reasons why the contract sum may change. These reasons should be specified in the contract, and may include the following:

- unforeseen work
- specialists' work
- agreed changes to design
- loss and expense to contractor
- liquidated damages
- changes in cost levels.

Unforeseen work

This falls into two categories:

1. Work which could not have been foreseen at the outset, such as unpredictable ground conditions, or unknown obstructions. It is a matter of interpretation of the contract who is liable for any additional costs incurred.
2. Work which was foreseen in qualitative terms, but the extent or precise quantity of which was unknown. An example is where the quantity of work cannot be accurately measured pre-contract and so appears in the bill of quantities as 'provisional'. This is often the case for substructure and drainage work.

Specialists' work

This derives from the traditional system whereby the client, through the architect, retains the right to nominate particular specialist subcontractors to carry out certain aspects of the works, such as services and special finishes. The work of nominated subcontractors is covered by prime cost sums in the bill of quantities. Since these reflect firm quotations, they should represent a reasonably accurate cost. However contractors often argue that they could control costs more effectively if they were free to choose their own subcontractors.

It is also worth noting that services form an increasingly high proportion of the total cost of buildings. The inability to measure accurately all services work at pre-tender stage can have serious implications for cost control.

Agreed changes to design

This is the area where costs may increase quite significantly. Throughout the course of the project, the client, or other members of the client's team, may have different thoughts on some aspect of design — layout, specification of finishes, and so on. Arguably, this behaviour is encouraged, because the standard contracts used with the traditional procurement system lay down rules on how to value variations, thus making it a self-fulfilling prophesy. For an agreed change to take place, the contract stipulates that instructions be given in writing. Although there are rules for valuing variations, in practice this does not always work smoothly. In any event, clients must accept that if they, or authorised members of their team, alter aspects of the design, then this is likely to have a price implication.

Loss and expense to contractor

Payments under this contractual provision are usually referred to as 'claims'. If a contractor believes that some action of the client has resulted in additional costs being incurred in carrying out the work agreed in the contract, then a justifiable claim for compensation exists. Although this differs from variations, there is an overlap. Variations are for additional work; loss and expense is for additional costs in carrying out the agreed works.

Typical reasons quoted by contractors to support claims include:

- failure to make the site available in time
- failure to provide adequate drawings and other information at the required time.

Loss and expense payments are intended to compensate contractors only for additional costs caused by some contractual failing on the part of the client. Therefore there is a considerable onus on the contractor to prove loss. This sometimes requires the services of 'claims consultants' and/or lawyers.

Liquidated damages

These arise when the contractor fails to complete the project by the time agreed in the contract. In this event the client reduces the payments made to the contractor. This is usually by an amount per unit of time late, for example £x per week. This amount is agreed and specified in the contract. Although this represents a reduction in the contract sum, it is not likely to represent a saving. This is because the sum agreed is meant to reflect a genuine pre-estimate of the losses which would accrue to the client as a result of late completion. The sum may be small for those clients where late completion is not of critical importance, but for clients waiting to open a new shop, the cost of delay through lost sales revenue could be quite considerable. Where the sum allowed in the contract for liquidated damages is not regarded as a genuine pre-estimate of loss to the client, then it is deemed to be a penalty, and the courts have the power to strike it down.

Changes in cost levels

In times of inflation, cost levels can change considerably during the course of a lengthy project. This raises the question of who is to bear this risk under the contract. If the contractor is required to bear the risk by having to submit a firm price tender, then this is likely to be reflected in pricing. If the client is to bear the risk, then the contractor can submit a tender based on current prices, but the contract will include a fluctuations clause which allows the contractor to be paid for certain increases in cost levels. The basis for the increased payment will probably be the formula agreed and accepted within the industry.

In summary

When the contract sum changes, it is quite possible that one of the parties will find its profit level affected. Much may depend on risk allocation within the contract. Of course, the effect of the payments chain should not be forgotten. Having considered profit, attention can now turn to liquidity.

Liquidity problems for the client

A construction project can be a rather lengthy process, and there may be a considerable time lag between paying out money and receiving a return. Land acquisition and construction costs have to be paid 'up front', while revenue only flows when the houses are sold or the office space is let. Furthermore, for a good part of the development process, the client will probably have to rely on expensive short-term finance. This is because less expensive long-term finance is only available when the asset it is being secured against (the building under construction) is nearing completion.

At feasibility stage, the client calculates that the project is likely to generate a certain pattern of cash flows which will eventually yield an acceptable rate of return. However, given the lengthy period between paying out and receiving back, a good deal can happen to disrupt the cash flow. This can have the effect of turning a potentially profitable project into a loss maker. Among the uncertainties which can disrupt the cash flow are:

- time taken to gain approvals
- changes in interest rates
- changes in demand
- construction difficulties.

Time taken to gain approvals

Before a project can be implemented, approvals such as planning permission must be obtained. As part of the feasibility process, the client calculates that a site can be transformed into a profitable project if, for example, a certain number of houses can be built on it, or a certain number of square metres of office space can be provided. The client may have a good idea of local planning policies. Even so, lengthy negotiations may be necessary before planning approval is granted, thus delaying the time when income will flow.

Changes in interest rates

Interest rates primarily affect the cost of finance. Their precise effect depends on the amount borrowed and the length of time involved. During the course of a project, interest rates can change several times. If rates are cut, then of course this can be an advantage. However, increases can have a severe impact on liquidity, affecting the financial success of the project, or even threatening the survival of the firm itself. Significant rate rises in the late 1980s and early 1990s had a particularly severe impact on many housebuilders and property developers who had expanded rapidly in response to the boom.

Changes in demand

Another consequence of a lengthy project period is that market conditions can change considerably. A housing or office development, started in boom conditions, can be severely affected by a recession. The consequent reduction in demand will lead to reduced prices and rents. This not only means lower profits, but also slower cash flow and extended requirements for short-term finance.

Construction difficulties

Technical and construction difficulties often occur in the course of a project. If these extend the construction period, then liquidity will be affected, because short-term finance will be required for a longer period. Which party bears the cost of this will depend on how risk is allocated in the contract. In the case where nobody is at fault, then the contractor may be entitled to an extension of time. This will affect the cash flow of both client and contractor. If one party is adjudged to be at fault, then contractual devices such as loss and expense, and liquidated damages come into effect.

Liquidity problems for the contractor

Some of the problems faced by the client are also applicable to the contractor, particularly construction difficulties. This depends on risk allocation in the contract. Changes in interest rates also have an impact, since these affect the cost of short-term finance.

The typical liquidity pattern which contractors may face is illustrated by the project cash flow diagram in Figure 22.2. The cash outflow is shown as S-shaped, indicating that payments are continuously made for labour, materials, subcontractors and so on, starting off slowly and then building up before tailing off again. By contrast, cash inflows are received monthly when the client makes a payment on account. This arrangement continues until practical completion, with final payment being made at the end of the defects liability period. Therefore, the cash inflow curve is shown as stepped. The precise position of each curve can vary, but it is frequently shown as in this diagram — that is, with the inflow curve sometimes above, and sometimes below, the outflow. Where inflow falls below outflow, this represents a need for short-term finance — 'a' and 'b' in Figure 22.3.

It is quite likely that the contractor will have to meet at least some short-term financing, if only because of the lengthy procedures involved between performing work and receiving payment for it. A typical chain of events is as follows:

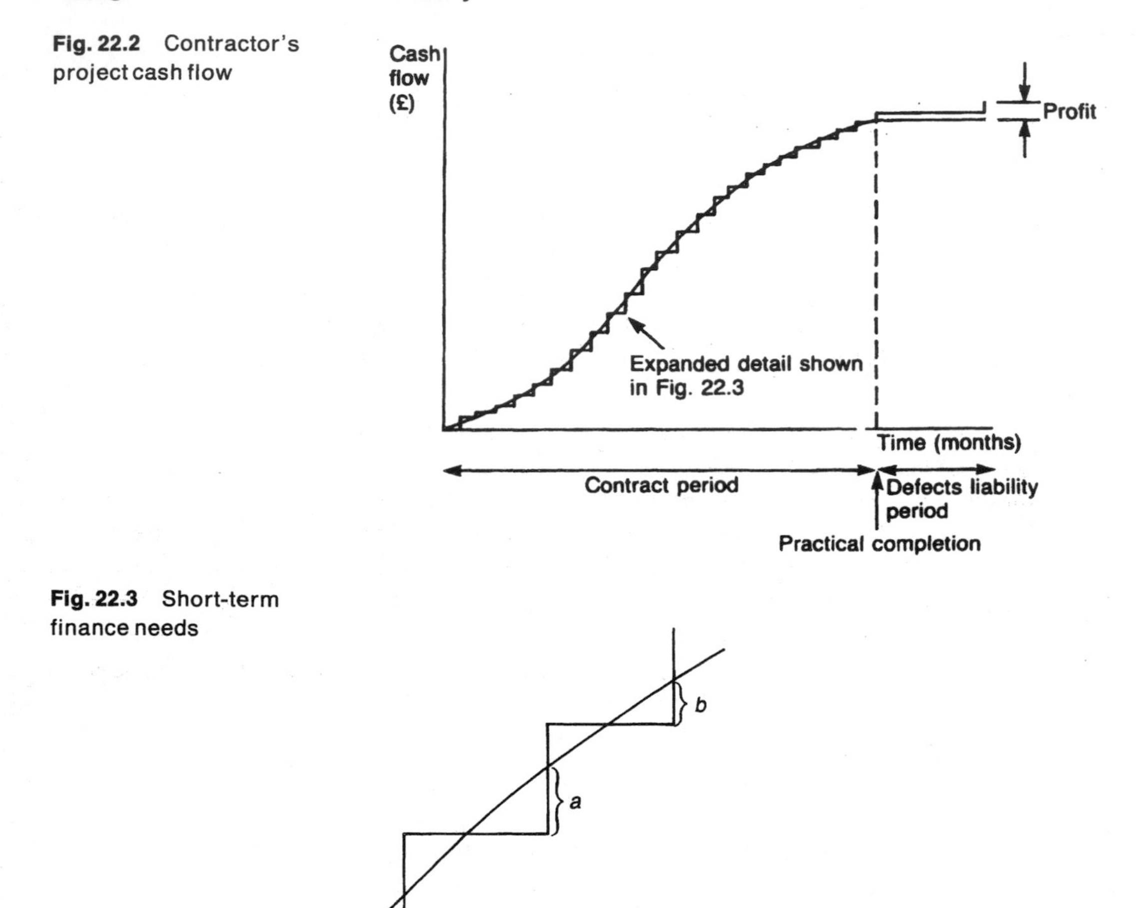

Fig. 22.2 Contractor's project cash flow

Fig. 22.3 Short-term finance needs

- contractor undertakes a portion of the works
- contractor values all work done on a regular basis
- valuation agreed with client's quantity surveyor
- valuation processed by client's quantity surveyor and sent to architect
- architect issues interim certificate and sends to client
- client pays.

Even if all runs smoothly, these procedures still take considerable time. If there are any hitches, the period is extended, with implications for the cost of short-term finance.

Dealing with liquidity problems

For any firm, the conventional ways of dealing with liquidity problems are by acquiring sources of short-term finance, such as bank overdrafts and trade credit. In the construction process, the existence of the payments chain allows for the use of trade credit. However, it can be argued that some parties exploit their position unfairly. In particular, contractors may be tempted to pass on their liquidity problems to subcontractors, by delaying payments as long as possible, and by insisting on 'pay when paid' clauses. Although this may improve cash flow, such an approach creates conflict.

However, there is a good argument for taking a different approach. By gaining a reputation for paying subcontractors promptly:

- the best will want to work for you
- prices quoted will not need to make provision for late payments
- preferential treatment may be received in times of shortages.

Apart from the above, additional solutions to liquidity problems for clients and contractors are also available.

Solutions for clients

Besides solutions such as bank and trade credit, clients of different types may use the following:

1. Where a client has several projects running concurrently, then cross-subsidisation may be possible. Projects with positive cash flows may be used to offset those with negative cash flows.
2. Housebuilders may sell 'off-plan', where the buyer pays a deposit at an early stage, in exchange for a guaranteed price.
3. Commercial developers may seek to find a tenant at an early stage of the project — a 'pre-let'. This makes the ultimate success of the project more likely, thus improving the chances of acquiring, at an earlier stage, long-term finance on more favourable terms.
4. Although selling assets in order to provide cash flow is not normally to be recommended, a change in market conditions may mean that surplus amounts of land are being held. These could be sold to release locked up capital.

Solutions for contractors

Bank and trade credit solutions are also commonplace, and, like clients, contractors may also be able to cross-subsidise projects in certain circumstances. As far as the individual project is concerned, much emphasis has been placed on the importance of contractors recording all the work they do on site. This information should then be promptly reported to the contractor's office, so that it can be established whether the work carried out is part of the agreed contract, or whether it is additional work with a financial implication. This is important for profit as well as cash flow. Among the procedures which contractors can adopt to improve cash flow are:

1. It should be ensured that interim certificates include all work carried out to date. This means providing all necessary information to the client's team so

that agreement can be reached on the value of variations and claims for loss and expense.

2. The time gap between certification and payment should be monitored. If this appears unreasonably long, then the reason must be ascertained. It may be due to excessively slow procedures or, worse, it may be an indication that the client has financial difficulties. In the worst case, work may be carried out for which payment is never received, because the client goes into liquidation.

In conclusion Many of the liquidity problems which can occur in a construction project have been considered from the viewpoint of the individual parties. However, it can be argued that these problems could be drastically reduced if the parties were more co-operative and less conflictual, a theme which will be taken up in the final chapter.

Summary

This chapter has considered various aspects of the financial control of construction projects. Financial control, as discussed in the context of this chapter, was first defined. The main focus was on the market transactions between the various parties involved in the project and how these affect profit and liquidity. An important feature of the construction process is the payments chain, where the extent of competition and market strength at each level is a critical factor. Following this, consideration was given to the various ways in which the cost to the client may alter in the course of construction. Finally, the particular problems of liquidity, as they affect clients and contractors, were considered together with some possible solutions to these problems.

Further reading

For a general overview of post-contract cost control, see Lavender (1990), application K. This topic closely links financial control to market transactions regulated through contracts. Several books on contracts in construction can usefully be referred to including Ramus (1989), chs 7–14; The Aqua Group (1990), chs 5, 6 and 7; Willis and Ashworth (1987), chs 15 and 16; and Murdoch and Hughes (1992), chs 11, 12 and 16.

The importance to contractors of recording all work carried out and reporting from site has featured prominently in papers from the Chartered Institute of Building, including Scott (1995), Bennett (1989) and Bell (1989).

23 Production management and construction

Introduction

The nature of production management generally was considered in some detail in Chapter 10. This chapter will start by reiterating some of the main elements of production management. One of the topics to emerge from this will be a study of how production methods have developed in the construction industry, compared with industry generally. This historical view will reveal some interesting comparisons. A major element of production management in construction is cost control through productivity performance. This differs from the type of cost control examined in the previous chapter, which considered cost control in the sense of prices paid for resources through market transactions. The various factors affecting productivity in construction will be considered in this chapter. Finally, some of the productivity techniques, such as work study will also be considered.

Main elements of production management

These were considered in Chapter 10, and can be summarised as follows:

1. Production creates the wealth which society requires by adding value to resource inputs.
2. The production process is presided over by the firm, which buys resources, uses them in production, and finally sells the resulting product.
3. Each product reaches the final customer after going through a number of stages in production, each presided over by a firm.
4. Efficiency in production is measured by productivity, which influences the costs of production, which in turn affects profits.
5. The production process is influenced by both technological and social factors.
6. Some improvements in productivity may be achieved without major expenditure on new resources, by altering production methods.
7. Sustained improvements in productivity usually require capital expenditure on equipment or training.
8. Although there have been variations in management strategy, the trend has

been towards greater mechanisation and division of labour, but with some degree of flexibility in recent years.

Development of production management in construction

As noted above, industrial production management has been dominated by scientific management approaches such as de-skilling, the division of labour, mechanisation and factorisation. This trend has continued, although amended by influences such as the human relations school of management and contemporary approaches such as core/periphery relationships, flexibility and subcontracting. The development of production methods in the construction industry has taken a slightly different path, as will be seen.

The development of production methods in construction can be described in four historical stages:

1. *Traditional*, until about the late 1950s.
2. *First industrial phase*, during the 1960s.
3. *Subcontracting phase*, during the 1970s.
4. *Second industrial phase*, since the 1980s.

This shows that industrialisation did not occur until quite late. Even then, it was not a complete success, and the process was interrupted. Each stage will be described in terms of its technological and social characteristics.

Traditional

In technological terms, traditional construction is characterised by site operations where skilled craft operatives work on relatively unprocessed materials. Skilled workers are supported by non-craft labourers, and a relatively small amount of plant. Therefore, a high proportion of the value of a building is added on-site, rather than off-site in a factory.

In social terms, craft workers tend to have a certain amount of autonomy. To a large extent they are free from direct management control. This means they can concentrate on achieving high quality and can take pride in what they are doing. From the management point of view, the craft operatives' attitude to their work may not fit with management's desire to increase productivity and cut costs.

When traditional construction was the norm, operatives were directly employed under a contract of employment. If these factors of autonomy and direct employment are taken together, it can be seen that the link between pay and productivity is fairly weak. Although the work may be of a high standard, the client may not be prepared to pay for it.

First industrial phase

Although some off-site prefabrication was used, traditional craft-based construction was still the norm well into the post-war period. Towards the end of the 1950s, however, greater attempts were made to emulate manufacturing. The construction process was industrialised by carrying out more of the construction process off-site in factories. A typical example from this period was systems building — wall and floor panels were prefabricated off-site, then bolted together on site. The intention was that large numbers of housing units could be built quickly, thus solving the housing shortage which had existed since 1945.

However, these buildings were not a great success, for the following reasons:

1. The main demand came from local authorities who required large numbers of inexpensive dwellings. However, the buildings were extremely unpopular with tenants because they were often high rise, damp and unpleasant.
2. Eventually the buildings were also regarded as unsafe, particularly in the aftermath of the Ronan Point disaster, where the whole side of a block of flats collapsed after a gas explosion blew out a wall panel.
3. Because of the subsequent reduction in demand, the economies of scale required for the economic production of the prefabricated units were no longer available.
4. Although the technological factors had changed, the social factors had not. The workforce was still organised along craft/non-craft lines, whereas what was required for this new technology was a semi-skilled workforce capable of accurately linking prefabricated panels and waterproofing joints.
5. Because the workforce had not been trained in the appropriate skills, more supervision was required. The work rate was slower than anticipated, and cost savings were reduced. Where there was insufficient supervision, the results were often poor.

Thus this first attempt to industrialise building production was largely unsuccessful — the systems were poorly designed and executed.

Subcontracting phase

Part of the reason for the failure of the first attempt to industrialise construction was the mis-match between the technological and social factors. Industrialised technology required a different type of workforce to that which existed. Consequently, changes to social and organisational factors gathered pace in the 1970s. The main change was the shift from direct employment to self-employment and labour-only subcontracting. The principal effect of this was to create a stronger link between pay and productivity. This was because those employed were now paid a lump sum for carrying out a set amount of work, rather than a regular weekly wage. Hence, this particular payments-by-results system came to be known as 'the lump'. Although many workers were self-employed, they were nevertheless often engaged by the same contractor for a considerable length of time. This gave rise to the legal problem of distinguishing between a contract of service and a contract for services, as discussed in detail in Chapter 18.

Apart from the social and organisational changes described above, technological changes also occurred. The demarcation lines between the various trades began to blur. Although traditional construction methods continued to be used, design tended to become simpler and more rationalised. In addition, many of the traditional materials were increasingly replaced by modern materials, which had a similar end result and appearance but were easier to work with. Examples include plasterboard, plumbing fittings, precast concrete, plastic pipes, etc.

Thus, by the end of the 1970s a new kind of workforce had emerged which was more flexible and multi-skilled, albeit at a level below that of the traditional craft operative. Subcontracting had become extremely widespread, and payment-by-results was common. Therefore, the need to earn a living wage seemed to become the main motivating factor. This was particularly the case because there was little job security for the self-employed.

Second industrial phase

With the changes described above, the stage was set for a new phase of industrialised construction. As noted, this had already started with the changes to materials, but was soon extended into a much wider use of factory-made components. These components were not specific to particular systems, as had been the case in the first industrialised phase. Instead their flexibility meant that they could be used by designers and constructors in a wide range of buildings. Examples of these types of component include:

- roof trusses
- doors and windows already hung in their frames
- sheet finishings
- cladding panels.

The effect was that a higher proportion of the value of a building was now being added off-site rather than on-site.

Once again, prefabricated components are being used structurally. For example, timber-framed houses are used in spite of some adverse publicity in the 1980s. Steel-framed office blocks were widely used in the commercial property boom of the late 1980s when clients were looking for fast completions. It is even possible to prefabricate whole rooms. For example, student halls of residence sometimes consist of completely fitted study bedroom modules. These are stacked, external walls and roof are added, and services are connected. During the commercial property boom, washroom and toilet 'pods' were prefabricated and lifted into place.

The impact of all these changes has been far reaching. The main focus of value added has shifted from the site to the factories where materials and components are made. Although the workforce is more flexible and has the potential to be multi-skilled, there has nevertheless been an increase in specialisation of a sort. This is due to the division, and possible further subdivision, of work into packages. For example, the plumbing subcontractor may further subdivide the plumbing package, and engage smaller sub-subcontractors. Each of these may undertake well-defined tasks such as:

- fitting appliances
- installing boiler equipment
- fixing pipes.

Each of these tasks is carried out repeatedly by a 'plumber', who specialises in one or other of these tasks.

The conclusion to be drawn is that, in a technological sense at least, building production has increasingly come to resemble manufacturing. Interestingly enough, the trend in manufacturing has been to adopt some of the social and organisational practices associated with the construction industry. For example, it was shown in Chapter 10 that contemporary approaches to production management in industry generally have included core/periphery relationships, flexibility of employment, and subcontracting or out-sourcing — all of which are features of the construction industry. A study of the history of production management, therefore, reveals that while construction has moved towards manufacturing in technological terms, manufacturing has moved towards construction in social and organisational terms.

Productivity in construction

To reiterate briefly what is meant by productivity:

1. Productivity measures efficiency — it is a purely physical non-monetary measure, relating output to resource input, usually labour.
2. Productivity influences the costs of production, which in turn affect profits.
3. Some improvements in productivity may be achieved without major expenditure on new resources, while sustained improvements in productivity usually require investment.

The main concern here is to examine those factors which affect productivity in a construction project; but first, some of the broader aspects which affect the industry generally will be considered. This is important because one of the problems associated with measuring productivity is the need to compare like with like.

Statements are sometimes made about productivity changes in the industry, which can be misleading. For example, the construction industry may have seen increases in productivity during the boom of the 1980s. This can easily happen in a boom because clients often require faster completions. Therefore they may be willing to pay more for faster, more mechanised, high productivity methods. This does not necessarily mean that productivity has been increased on a sustainable basis. Conversely, overall productivity statistics may worsen in a recession.

However, this may not be an indication of a serious decline; it may simply be that clients are not investing in new building, but are instead refurbishing existing buildings. Refurbishment work is often labour intensive, and therefore productivity, as measured by output per unit of labour input, will be lower. Having noted the above, it must still be said that recessions often do lead to a weakening of underlying productivity conditions. This is because regular programmes of investment, in capital equipment and skills training, are often neglected at such times.

The relationship between productivity and profit margins is often debated. When margins are low, it could be logically argued that productivity suffers because neither the funds nor the confidence to invest exists. This should mean that when profit margins increase, there should be an increase in investment and, hence, productivity. However, there is no guarantee this will happen. Instead, higher profits could lead to higher dividends being paid to shareholders, and/or to increased merger and takeover activity. This is often regarded as a problem for UK industry generally, not just construction, and may derive from the way the capital markets operate.

Turning now to the factors which affect productivity in construction projects, these are, of course, numerous. Horner (1982) made a list of ten:

- quality, number and balance of labour force
- motivation of labour force
- degree of mechanisation
- continuity of work
- complexity of project
- required quality of finished work
- method of construction
- type of contract
- quality and number of managers
- weather.

Arguably the list is not complete, and there are overlaps, but it forms a useful framework for discussion.

Quality, number and balance of labour force

This is a very wide-ranging category. Quality of the labour force relates to skill, either intrinsic or acquired through training. Sometimes a high level of skill may detract from productivity, for example, where the worker is anxious to produce work of a higher quality than that required by the specification. But generally a highly skilled person is able to carry out a high quality of work at speed. The skills base of the workforce tends to diminish in a recession as less training takes place. This tendency always causes a supply problem when an upturn in the economy occurs. Furthermore, the trend towards subcontracting has meant that fewer firms are willing to train, especially when competition is so intense. Inevitably, this represents a problem for the whole industry.

Number and balance are quantitative factors. Having too many workers on site could be detrimental to productivity due to logistical problems. The question of balance occurs within and between trades. There should be appropriate numbers of workers at each level of skill. Given that various activities fit together into the project programme, as discussed in Chapter 21, it is important that there are an appropriate number of gangs in each trade on site, otherwise there will be more 'idle' time, thus reducing productivity.

Motivation of labour force

Motivation was discussed in some detail in Chapter 15. There have been several studies which have related some of these general ideas to the construction industry — for example, Nicholls and Langford (1987), who examined the motivation of site engineers in relation to Herzberg's two-factor theory. While these human relations theories have often been applied to non-manual labour, it has usually been assumed that manual labour responds better to scientific management type approaches. This has been taken to mean financial incentives such as bonus schemes. However, this may not be altogether true. It has already been argued that craft workers often take great pride in their work — more than clients are sometimes prepared to pay for. The idea is also established in economic theory that workers work less hard if pay rises. This is because they have some 'target earnings' which they wish to achieve, and if this level can still be achieved by working fewer hours or less hard, then workers will choose this option. Much of this discussion becomes less relevant when workers are self-employed and paid by results. This is especially the case when there is high unemployment and pay rates are low. At such times, considerable effort is required just to secure a living wage.

Degree of mechanisation

If machines are used instead of labour, then productivity (in arithmetical terms) will automatically rise, simply because of the way that productivity is measured. This, of course, highlights the potential shortcomings of productivity data. An apparent improvement may in fact conceal a not particularly efficient use of an inappropriate machine! So although there may have been a nominal productivity gain, the high price paid for the machine may have greatly increased costs. Having said this, it should be emphasised that the main purpose of using machines is to increase productivity and gain a time and/or cost advantage. As noted before, changes in construction methods towards the greater use of components have tended to increase mechanisation.

Continuity of work

As stated above, a project programme will have been drawn up in order to make the best use of resources, and therefore with productivity in mind. Any break or lack of continuity in the programme is likely to adversely affect productivity. Horner lists four factors which can cause a lack of continuity. These are:

- problems with supply of materials
- performance of other contractors or subcontractors
- availability and adequacy of technical information
- variations.

The latter two factors were discussed in the previous chapter, in the context of alterations to the contract sum. For example, the contractor may be entitled to extra payments in certain circumstances, such as where there is a break in the continuity of work. It has also been noted that because most work is divided into work packages and subcontracted out, the main contractor has less direct control over the whole process. Problems of co-ordination exist which could easily disrupt the works.

The supply of materials, and their flow around the place of production, has received a good deal of attention. The just-in-time approach, first discussed in Chapter 10, has been applied to construction — see, for example, Baxter and Macfarlane (1992).

Complexity of project

Complexity can affect productivity in several ways. The more complex a project, the more difficult it will be to plan, and hence the greater the likelihood of continuity problems. Furthermore, complexity may discourage the use of mechanisation, perhaps because of difficulties in moving around the site, or because there is not enough similar work to achieve economic plant usage.

Required quality of finished work

It might be expected that achieving higher quality may take longer and therefore reduce productivity. To an extent, this may be true for traditional construction, but the use of modern factory-made components means that the 'quality' is often in the materials, and a skilled worker can install them with no particular loss in productivity. The approach of total quality management, discussed in Chapters 10 and 20, involves designing the production process at the appropriate level of quality to satisfy customer needs, so that everything is right first time. In this respect high quality should not mean an undue sacrifice of productivity.

Method of construction

With reference to the historical survey of the development of construction methods earlier in this chapter, it will be apparent that the quest for higher productivity has stimulated change. This is because higher productivity should lead to reduced costs, which in turn should lead to higher profits. It was shown that construction methods have evolved — a higher proportion of value is now added off-site through the use of prefabricated components. All things being equal, this should help increase productivity on site, as there are fewer things to go wrong.

Although technological changes have led to more efficient production methods, and therefore higher productivity, this may be off-set by organisational changes. For example, the change to large numbers of subcontracted work packages, and the accompanying contractual complexity they create, can produce problems which damage productivity.

Type of contract

The type of contract and procurement method may not directly affect productivity. However, it should be noted that certain procurement methods tend to be used in particular circumstances. For example, construction management is a procurement method designed for fast completion and close client involvement. However, since a large number of work packages tend to be used, the productivity problems referred to in the previous paragraph become apparent. When using procurement methods under single point responsibility, there is less likelihood of the time allowed being extended. This puts a greater onus on the contractor to achieve higher productivity.

Quality and number of managers

This will depend on the way the project is organised, and the supply of competent managers. Under traditional construction methods, where the whole work was under the direct control of the main contractor, a large number of supervisory staff were usually employed — general foreman, various trades chargehands, and so on. Where most of the work is subcontracted out, there will be fewer supervisory staff employed by the main contractor. In such cases, supervision will depend much more on how the subcontractors supervise their own work. In theory, subcontractors have the financial incentive to use high-productivity methods, as they are being paid a lump sum to complete a work package by a specified time. However, not all subcontractors have the necessary management expertise.

The supply of competent managers is determined by the numbers entering the industry, and the quality of training. Unfortunately, training is often neglected during recessions. Moreover, good people may leave the industry due to poor career prospects. This leaves a 'managerial gap' when the economy takes an upturn. Inadequate management cannot be good for productivity.

Weather

Bad weather can have a detrimental impact on the construction programme and, hence, hinder productivity. This will clearly be more of a problem during groundworks and before the building is weathertight. It is likely that modern construction methods — using more mechanisation, the greater use of factory-made components, and less wet trade work — have reduced the scale of the weather problem. However, a construction site is not an enclosed factory.

In summary

There are a large number of factors which can affect productivity in construction. Perhaps the issues which arise most frequently can be better understood by referring to the main conclusions of the previously discussed historical summary:

1. In *technological* terms, there has been a move towards the greater use of factory-made components, meaning that more of the value of a building is added off-site.
2. In *social and organisational* terms, the work is divided into work packages, most of which is subcontracted out and ultimately carried out by self-employed workers on payment-by-results systems. This is more flexible but gives rise to control and co-ordination problems.

Productivity techniques

Since the days of F.W. Taylor, supporters of the scientific management school of thought have believed that it is possible to determine the one best way to carry out a work task, and hence achieve maximum productivity. Techniques for studying productivity and implementing improvements have therefore been developed. Some of these, such as operations research, take full advantage of statistical techniques and the use of computers — see Baxendale (1984). However, the best-known collection of techniques is known as **work study**. This comprises:

- **method study**, which analyses the actual tasks in order to improve the ways in which they might be done
- **work measurement**, which calculates standard times within which tasks should be carried out, thus establishing standards for control.

Together, the use of these techniques should result in higher productivity which benefits:

- the employer, through lower costs and higher profits and/or
- the workers, through higher wages and better conditions.

These studies can generate an enormous amount of data which can be used in production management. If used throughout the organisation, this data can also have a number of other applications, for example:

- estimating
- planning
- payments systems, such as piecework rates and incentive schemes.

Before briefly considering the main elements of work study, it is worth stating some reservations as far as their use in construction is concerned. To set up a work study system requires a considerable expenditure of resources. When production was directly controlled by the main contractor, the resources may have been available. However, with the work now divided into packages, it is less likely that the subcontractors, who now control production, will have sufficient resources available. In any event, some may argue that work study is too production orientated, and that more market-orientated approaches such as total quality management are more appropriate for the construction industry.

Method study

As stated above, this seeks to find the best way of carrying out a particular task. All resources are looked at, including the use of labour, materials and plant. In addition, site layouts, flow of materials, etc., are also examined. The process of method study begins with the selection of a particular activity which is usually repetitive to enable future savings to be made. A number of stages in the process will then unfold:

1. *Record*, where the activity being studied is detailed. A variety of charts and diagrams are available according to the type of activity — see Forster (1989), ch. 8, for a description.
2. *Examine*, where the activity is critically analysed. This involves identifying the problems with the way the activity is currently being carried out, and assessing the alternatives.
3. *Develop*, where, as a result of the above critical examination, the best method is developed.
4. *Install*, where the better method is explained to those who have to make it work. The co-operation of supervisors and operatives will be easier to achieve if they have been involved in the previous stages.
5. *Maintain*, where the installed system is checked in operation to make sure it is functioning properly, and to make adjustments if necessary.

Work measurement

This is the process by which control standards are set, and is a similar concept to the standard costing techniques discussed in Chapter 11. Indeed, the information gathered by work measurement can be used when formulating standard costing and budgetary control systems. The purpose of work measurement is to determine how long it should take an average worker to perform a certain task. The information is particularly useful for setting bonus rates or payment-by-results rates. The process of work measurement has similarities with method study, as the same order — select, record, examine — is also followed. However, the core of the process is measure, for which the following techniques are available:

1. *Time studies*, which involve the physical measuring of the time taken to carry out a task using a stop watch. This is to set the standards.
2. *Activity sampling*, which checks that the work is being carried out according to plan. This is therefore an important control mechanism.
3. *Synthesis*, which involves obtaining times or durations for a task, based on data which already exists for other activities. This is therefore a composite item.

4. *Analytical estimating*, which is used for tasks of an unusual or one-off nature for which data does not exist, or which cannot be easily measured. To arrive at an estimate, previously recorded times will be used where possible, with an assessment made of the rest.

As previously mentioned, work study may be used for purposes such as setting bonus targets. They aim to be objective and thus fair. The gains resulting from better productivity, obtained through work study, can be divided between employer and workers in agreed proportions.

Summary

This chapter has focused particularly on productivity performance in construction. The first requirement was to summarise the main elements of production management, as discussed in Chapter 10, thus laying the foundations for this chapter. A historical survey of the development of construction methods was compared with a similar survey of the development of production methods in industry generally. Changes in production methods are usually stimulated by the quest for better productivity, this being a key to lower costs and, hence, higher profits. Therefore, some of the factors which influence productivity in construction were examined. Finally, brief consideration was given to techniques such as work study, which are used to ensure that tasks are carried out more efficiently.

Further reading

For an economics based view of production methods in construction, including their historical development, see Lavender (1990), application M. For productivity in construction, see d'Arcy (1993) and Horner (1982). For the application of just-in-time to construction, see Baxter and Macfarlane (1992). For motivation of construction site staff, see Nicholls and Langford (1987).

For productivity techniques, see the following texts for more detail on work study and incentive schemes: Forster (1989), ch. 8; Calvert *et al.* (1995), ch. 20; and Baxendale (1984), who discusses the use of operations research on site.

24 Future prospects for the management of construction

Introduction

Before anyone can deal with the future, they need to be able to forecast what is going to happen and plan accordingly. This does not mean simply passively changing with the wind, but rather a more proactive approach involving deciding where one wants to be, responding to the circumstances, or even creating the circumstances which will enable the desired results to be achieved. For the organisation, this means developing the ability to manage change — so that, as a minimum, the firm will survive, but, more ambitiously, so that it will be able to thrive in response to change, and even create the change.

To a great extent, the management of change has been a theme of this text. For example, much has been made of the market-orientated firm, which puts the needs of customers at the centre of its activities. This does not mean simply giving customers what they say they want, it also means developing ideas, goods and services, from which the customers are likely to derive benefit. Similar issues apply in government — central and local. Good government is not just about responding to what people appear to want, in the hope of being re-elected — it is also about marking out a vision for the future of society.

The management of change is a topic in its own right. This chapter will discuss some of the main principles, with reference to other writings and to the previous chapters in this text. For managing change in construction, or any other industry, it is important to be aware of likely changes in market conditions, including shifts in demand and the structure of the industry.

Another important change could be the way in which an industry should conduct its affairs. There has been much discussion over the years about contractual relations in the construction industry, and the damaging effect of conflict. The debate has been intensified in the wake of the report by Sir Michael Latham, *Constructing the Team*, published in 1994.

In addition to market conditions and the internal workings of the industry, there are numerous factors which could have a significant impact on the future management of construction. Prominent among these are concerns about the environment, which include issues such as planning policies and infrastructure provision.

Management of change

Many of the chapters in this book have been concerned with the management of change. For example:

1. Chapter 2 showed how the dominant schools of management thought have changed over time in response to economic, social and political conditions.
2. Chapter 3 explained, as one of Drucker's five survival objectives, that a firm must be adaptable to change.
3. Chapter 4 pointed out the difficulties firms have in responding to changes in public policy.
4. Chapter 7 discussed why some organisational structures are more adaptable to change than others.
5. Chapter 9 showed that market-orientated firms have to adapt their marketing approach when necessary.
6. Chapter 10 discussed how production management has become more flexible in many organisations, with the use of just-in-time and similar ideas.
7. Chapters 13–16 considered variations in different aspects of human behaviour within the organisation. Some management and leadership styles, motivation strategies and group structures were shown to be more adaptable to change than others.
8. Chapters 19–23 examined various aspects of management related to the construction firm and the construction project. Many of these aspects emphasised change and some will be further referred to in the course of this chapter.

Change is often assumed to be a one-way process. For example, it might be supposed that the development of a new piece of plant, such as a pump to transport concrete to the upper floors of high-rise buildings, will change production methods for ever. Similarly, if a particular construction method, such as timber-framed housing, were to lose favour, then it might be expected that this would be a permanent change. Some change is one-way, or linear, but other kinds of change may actually be circular, leading back in some respects to what has gone before. Therefore, to further examine the management of change, linear and circular change will be considered separately, although it should be noted that many actual changes may be a mix of linear and circular.

Linear change

When analysing influences for change in organisations, one method is to classify them into:

- external influences
- internal influences.

External influences

The main influences, discussed in Chapter 4, were market pressures and public policy. To these can be added technological change. Some of these changes are of a short-term nature, such as fluctuations in the market or shifts in interest rate policy. However, other changes are of a long-term, or structural nature. For example:

1. Many markets have changed to become more global in character. Many domestic markets have become increasingly dominated by large firms, causing medium-sized firms to adapt their market position quite significantly.
2. Public policy has changed decisively, away from public spending and ownership towards private spending and privatisation. This has necessitated a major rethink for those firms who placed great reliance on obtaining public sector work.
3. Technological change has continued, stimulated by factors such as advances in electronic controls and information technology. The changes have been more dramatic in some industries than in others.

Internal influences

As discussed in Chapter 5 and subsequently, these derive from the people in an organisation. Again changes may be due to short-term factors, such as availability of certain kinds of labour, while other changes are longer term. Examples may be a change of ownership or senior management, which introduces a new organisational structure, as discussed in Chapter 7, and/or a new culture, management and leadership style, as discussed in Chapters 13 and 14. When management wishes to introduce changes of this kind, there could be difficulties, depending on the nature of the change. To gain the support of the existing employees and customers may require skilful management of the change.

Circular change

This term can be used in the current context to also include the case where forces are operating in different directions — some for change, some to prevent change. A similar idea was explained in the context of marketing management, in Chapter 9. In seeking to promote its products, a firm must raise the awareness of potential customers through a number of stages — from being unaware of the product, through to purchase. However, there are always countervailing forces, such as the elapse of time, which tend to reverse the process. A similar situation can occur in the management of change. A new senior management team may wish to make radical changes in the way the firm operates, but may meet resistance from the existing middle management team. One way of formulating this situation was developed by Lewin (1951) in his Force-field theory. This suggests that there will often be resistance to change, which can be best overcome by consultation and gaining acceptance for the proposed changes.

As previously mentioned, not all changes turn out to be permanent. Furthermore, many so-called 'changes' may in fact be a return to a previous norm, or may at least have elements of retrenchment. Some examples are as follows:

1. The move towards industrialised building in the 1950s, as described in Chapter 23, appeared to be permanent but failed for a number of reasons.
2. After adverse publicity in the 1980s, timber-framed housing in the UK appeared to have been dealt a fatal blow, but has returned in recent years.
3. The move towards free markets, including flexible labour markets, is not a new innovation dating from the 1980s, but a return to pre-1939 practice.
4. Working from home is not a new phenomenon; indeed, it has long been common practice in industries such as clothing. What is new is the remote working possibilities created by developments in information technology.

The conclusion which can be drawn is that although there may be forces stimulating a continuous linear change, there are also forces resisting change. There is also the tendency that 'history repeats itself' — practices which seem to have disappeared, return, albeit in a different guise.

Considering the UK economy as a whole, the period after 1945 witnessed a move away from the traditional Anglo-Saxon model of free markets, towards a more European interventionist model. This was widely regarded as a permanent change, particularly as both major political parties supported it. However, the emphasis has now changed and in many respects the old orthodoxies have been re-established. It could be that the UK never really did adopt the European model wholeheartedly, and although government intervention and public spending did increase, many Anglo-Saxon institutions such as financial, land and property markets were left untouched. For a critique of, in particular, the financial markets, as compared with European approaches, see Hutton (1995b). The position of the UK economy does create ambiguities. Although part of the European Union, the government seems ill-at-ease with many of the ways in which the other members conduct their affairs.

For the remainder of this chapter some of these aspects of change, as they relate to construction, will be further examined.

Changes in the construction market

This topic has arisen several times in this text. Chapter 9, on marketing management, discussed the importance of the firm being market orientated, while Chapter 19 pursued this by considering market strategies for construction firms. Numerous changes have occurred which have required the construction firm to be adaptable. Some of the examples discussed were as follows:

1. There has been a change in industrial structure, with the emergence of a relatively small number of large diversified national contractors who

compete for a range of work. This has made it difficult for medium-sized regional contractors to sustain their position, and many have had to expand, merge or become subcontractors.

2. Public sector demand has fallen, thus requiring contractors who relied on such work to adopt a more varied market strategy. Although public sector work used to fluctuate regularly in accordance with public capital spending programmes, arguably private sector demand, associated with, say, housing and commercial property markets, is even more volatile.
3. Private sector clients have become more professional in their approach, requiring an appropriate response from contractors.

In looking forwards to the latter years of the 1990s, construction firms will need to assess where areas of market opportunity are likely to be. A few pointers may be as follows:

1. It is unlikely that the housing market will reach the heights of the late 1980s. Having said that, buyers often show a preference for new houses, and housebuilders will need to market their products carefully. For example, there may be more emphasis on helping buyers make the purchase by taking their existing house in part exchange, thus breaking any buyers' chain.
2. The commercial property market for offices seems over-supplied in most urban areas, but there could be strong demand for refurbishment.
3. There are unlikely to be significant public funds for infrastructure provision, no matter how necessary this may appear to be. The government is apparently hoping for private funding of these schemes. This does offer a new opportunity for construction firms, but one which carries major investment obligations and considerable risk.
4. Recent years have seen great diversity in procurement methods on offer, in addition to the traditional system of selective tendering. At some point the situation may stabilise and a new norm may emerge. It is possible that firms who are able to offer clients single point responsibility, or who are willing to enter into partnering arrangements, will be in a stronger market position.

The conduct of the industry

The last section was substantially concerned with how individual organisations need to be ready to adapt in the face of changing circumstances. However, there is also the overriding issue of how the industry as a whole behaves. If this is unacceptable — for example, in terms of achieving cost, time and quality targets — then fewer clients will wish to undertake construction work and the whole industry will suffer. Arguably, many of the problems arise from the array of market relationships in the industry, each of which is regulated by a commercial contract. This phenomenon has been characterised in this text as the payments chain. As explained in Chapter 22, the operation of the chain can lead to conflict, increasing costs and liquidity problems. The effects of the chain are exacerbated by the fact that unequal competition exists, with some firms having monopoly power over others. When the industry is in recession, competition becomes even more intense, accentuating the effects of the payments chain even further.

Much comment has been made over the years about this state of affairs. There have been numerous official reports recommending change but thus far the problems persist. As stated in the introduction to this chapter, the Latham Report, published in 1994, has generated much debate. There are many proposals contained in this report, which can be grouped under the following headings:

- role of the parties
- contracts
- cost savings.

Role of the parties

All parties to the project are mentioned — clients, consultants, contractors, subcontractors. Special importance is given to the client: 'Implementation begins with the client.' Many private sector clients are expert, either as professional developers or as regular commissioners of construction work for their own organisation. It is suggested that they form a Construction Clients' Forum. In addition, a guide to briefing for clients should be drawn up for those who lack their own wide experience. There should also be codes of procedure for matters such as the selection of consultants, contractors and subcontractors.

The role of the public sector, particularly the Department of the Environment (DoE), is very important both as facilitator and sponsor of legislation and as a 'best practice' client.

Contracts

As previously stated, contracts, both their setting-up and operation, have long been problematical in the construction industry. The report has a good deal to say on these matters, including:

1. Partnering, that is, long-term relationships between the various parties, should be encouraged.
2. A family of interlocking standard contracts should be used based on, what was then, the New Engineering Contract (NEC).
3. There should be legislation in the form of a Construction Contracts Bill to give statutory backing to the standard forms of contract.
4. The Bill should also outlaw certain unfair contract practices, such as 'pay when paid' clauses. These were described in Chapter 22 of this text as one of the aspects of the payments chain particularly damaging to subcontractors.
5. Mandatory trust funds should be set up to clearly define and protect

moneys such as retention, which is being held temporarily by one party on behalf of another.

6. Adjudication should become the normal method of dispute resolution.

Cost savings

One of the more eye-catching aspects of the report was the stated target of 30 per cent cost savings which should be achievable by the year 2000. This means that a building of a given specification should be buildable for less. Cost control has featured strongly in this text. For example, it was explained in Chapter 22 that costs generally derive from:

- the price paid for resources acquired through market transactions, and subject to a commercial contract
- the efficiency with which those resources are used, as measured by productivity.

Chapter 22 considered the financial implications of market, and therefore contractual, relationships, while Chapter 23 concentrated on productivity matters. Chapter 21, which includes cost implications of time, is also relevant.

To achieve any cost savings, let alone 30 per cent, consideration needs to be given to market transactions and productivity. As for market transactions, the previous two sections on the role of the parties and on contracts are highly relevant. After all, many of the problems of cost and time over-runs stem from contractual problems and the failure of parties to the project to perform adequately. Therefore some of the cost savings should be achievable from such measures as:

- the use of partnering and/or single point responsibility
- careful selection of consultants and contractors
- matching design more closely to client needs
- allowing adequate time and resources for developing brief, design and procurement strategy
- developing better forms of contract and outlawing unfair contract terms.

As mentioned above, productivity was discussed in Chapter 23. As shown, there has been increased use of mechanisation and factory-made prefabricated components. To increase cost savings further, these trends should be continued. In addition, consideration should be given to:

- improving training and motivation of management and operatives
- more standardisation in design
- gathering better data on which to plan and control projects
- making appropriate use of information technology
- improving the flow of project information
- giving consideration to the application of just-in-time through better planning and relationships with suppliers.

The debate about *Constructing the Team* will undoubtedly continue, and many believe the results will be critical for the future of the industry. As recommendation 1 of the report states: 'Previous reports on the construction industry have either been implemented incompletely, or the problems have persisted. The opportunity which exists now must not be missed.'

Environmental concerns

The last section of this chapter will discuss some broader issues impacting on the industry. Almost everything which the construction industry does affects the environment. In fact, management of construction is often now regarded as part of the broader study of the built environment. There is greater public awareness and concern about environmental matters, and this is likely to be reflected in public policy. Therefore, it is expected that people as consumers, or as decision makers in organisations, will affect demand for construction work. Similarly, public policy decisions will affect both public sector demand and the legal and economic framework within which private sector demand takes place.

Under this heading, a number of issues will be briefly considered for their effect on construction, including:

- planning policies
- infrastructure provision
- production methods and energy usage.

Planning policies

These determine the types and scale of development allowable on given sites; therefore they are primarily concerned with land use which is a key environmental issue. The obvious effect on construction is that what will be allowed must be taken into account at the feasibility stage of a project. For example, the number of houses or the area of office space that can be built on a site will directly affect the potential revenue that can be generated. Of course, this is always the case, but, in the context of this chapter, it is changes in planning policy which construction organisations will need to adapt to. Some examples of policy changes which can affect construction are:

- no longer allowing new housing in a particular area, but encouraging refurbishment instead
- refusing permission for further out of town shopping centres
- restricting future office development to, say, four storeys in height.

All these affect the shape of demand for construction work. For example, in the 1980s there was a huge expansion of retail facilities on out-of-town sites. The large supermarket chains had very substantial construction programmes, providing a high proportion of the workload of many consultants and contractors. Although some store building has continued, this policy has now been questioned, and many local authorities are trying to prevent similar developments in the future.

Infrastructure provision

Traditionally, much of the responsibility for providing infrastructure has rested with the public sector. Many contractors have grown large on the basis of road construction and other civil engineering works. Two major changes have been taking place:

1. The government no longer sees itself as the main provider of infrastructure. It wishes to rely much more on private finance, through privatised utilities and involvement by financial institutions, contractors and others.
2. There is a growing feeling that building more roads to enable ever-increasing numbers of car journeys is not the answer to transport provision.

These changes will require different approaches by organisations in the construction industry. The potential for involvement in consortia providing new railway schemes, is one possibility.

Production methods and energy usage

This obviously includes work on site and the impact it has on the environment in terms of creating traffic congestion, noise and pollution. In addition, there are the effects created off-site. This is of increasing importance since a higher proportion of the value of buildings derives from factory-made components. Therefore attention must also be paid to the production of components and the impact this has on the environment. Finally, the efficiency of buildings in use is important, with matters such as energy savings and life cycle costs of great interest to house purchasers, commercial and public sector clients.

These are some of the many ways in which the construction industry affects the environment. Construction organisations will need to address public concerns. They must reflect those concerns in the products and services they offer and in the way they market them.

Summary

This final chapter has looked ahead and given some pointers to the changes likely to affect the management of construction, and how construction organisations might respond. The management of change is an important topic in its own right. To some extent this has been a theme throughout the text, and some of the general principles were reiterated in this chapter.

When specifically considering construction, it is important to assess likely changes in market conditions, since a response is an essential element in the strategy of market-orientated firms. The construction industry as a whole has a poor image and does not have a good reputation for consistently delivering projects within the prescribed cost, time and quality requirements. The report *Constructing the Team,* published in 1994, has stimulated a debate on how the industry might remedy its problems and make cost savings by implementing improvements in contractual arrangements. Finally, some of the public concerns about the environment were considered, indicating areas which need to be addressed by construction organisations. Thus the successful management of construction is concerned with the future as well as the present.

Further reading

Some ideas for further reading have been given in the text. For general overviews on the management of change, see Fryer (1990), ch. 9, and Cole (1993), ch. 26. For current information, see journals such as *Building* and the quality press.

Bibliography

Argyris C 1960 *Understanding organisational behaviour,* Dorsey.

Argyris C 1964 *Integrating the individual and the organisation,* Wiley.

Aqua Group 1990 *Contract administration for the building team,* BSP.

Ball M 1988 *Rebuilding construction — economic change and the British construction industry,* Routledge.

Barrie G 1994 10 Sweett chiefs bag £6.25m. *Building* 16 December: p. 5.

Barrie G 1995 Home truths. *Building* 23 June: p. 20–2.

Baxendale AT 1984 *Use of operations research on site,* CIOB Technical Information Service, No. 36.

Baxendale AT 1992 Integration of time and cost control, CIOB Construction Papers, No. 7.

Baxter LF and Macfarlane AW 1992 Just In Time for the construction industry, CIOB Construction Papers, No. 14.

Berle A and Means G 1932 *The modern corporation and private property,* Macmillan.

Belbin RM 1981 *Management teams,* Heinemann.

Bell GD 1989 *Financial management in construction,* CIOB Technical Information Service, No. 108.

Bennett JF 1989 *The contractor's quantity surveyor and reporting from site,* CIOB Technical Information Service, No. 106.

Bingham T 1994 When self-employed means employee. *Building* 27 May: pp. 26–7.

Birkbeck D *et al.* 1995 Structural shift — the future of contracting. *Building* 26 May: pp. 29–47.

Blackburn R 1972 The new capitalism. In Blackburn R (ed), *Ideology in Social Science,* Fontana.

Braverman H 1974 *Labour and monopoly capital — The degradation of work in the twentieth century,* Monthly Review Press.

Calvert RE, Bailey G and Coles D 1995 *Introduction to building management,* Butterworth.

Chevin D 1993 Perfect pitch. *Building* 4 June: p. 45.

Chevin D 1994 Life on the margins. *Building* 9 September: pp. 22–3.

CIOB 1992 *Code of practice for project management for construction and development,* CIOB.

Clarke L 1992 *Building capitalism — Historical change and the labour process in the production of the built environment,* Routledge.

Cole GA 1993 *Management: Theory and practice,* DP Publications Ltd.

Cole J 1990 The choice of construction planning systems. *Chartered Builder* September/October: pp. 11–13.

Cooke B 1992 *Contract planning and contractual procedures*, Macmillan.
Cooper P, Barrick A and Stewart A 1994 Time to cull. *Building* 21 January: pp. 18–20.
Cowe R 1995 Tarmac melts out of house building. *The Guardian* 4 August: p. 15.
Cyert R and March J 1963 *A behavioural theory of the firm*, Prentice Hall.
d'Arcy J 1993 Productivity: The Shame and the Sham. *Contract Journal* 12 August: pp. 12–13.
Davies DB 1992 *The art of managing finance*, McGraw-Hill.
Dixon R 1991 *Management theory and practice*, Made Simple series.
Dow R 1995 Risky Business? *New Builder* 19 May: pp. 27–8.
Drucker PF 1972 *Technology, management, and society*, Pan.
Edwards R 1979 *Contested terrain — The transformation of the work place in the twentieth century*, Heinemann.
Eldridge N 1990 How often should I change my corporate identity? *Chartered Builder*. November: pp. 12–13.
Farnham D and Pimlott J 1995 *Understanding industrial Relations*, Cassell.
Fayol H 1949 *General and industrial management*, Pitman.
Fiedler FE 1967 *A theory of leadership effectiveness*, McGraw-Hill.
Fisher N 1986 *Marketing for the construction industry*, Longman.
Fisher N 1991 Marketing. *Chartered Builder* April: pp. 14–15.
Forster G 1989 *Construction site studies — Production, administration and personnel*, Longman.
France G 1993 Plan it sweet. *Building* 10 September: p. 58.
Friedman AL 1977 *Industry and labour — class struggle at work and monopoly capitalism*, Macmillan.
Fryer B 1990 *The practice of construction management*, BSP.
Galbraith A and Stockdale M 1993 *Building and land management law for students*, Newnes.
Galbraith JK 1972 *The New Industrial State*, André Deutsch.
Gouldner A 1955 *Patterns of industrial bureaucracy*, Routledge & Kegan Paul.
Graham HT and Bennett R 1992 *Human resources management*, M&E.
Hamilton B 1990a Key characteristics. *New Builder* 21 June: pp. 26–7.
Hamilton B 1990b Relationship choices. *New Builder* 28 June: pp. 18–19.
Hamilton B 1990c Outlining concepts. *New Builder* 5 July: p. 8.
Hamilton B 1990d Time is money. *New Builder* 12 July: pp. 24–5.
Hamilton B 1990e Design direction. *New Builder* 19 July: p. 25.
Hamilton B 1990f A team effort. *New Builder* 26 July: p. 18.
Handy C 1993 *Understanding organisations*, Penguin.
Harper K 1994 Thatcher legacy changed industrial landscape. *The Guardian* 24 May.
Harris F and McCaffer R 1991 *The management of contractor's plant*, CIOB Technical Information Service, No. 127.
Harrison RS 1993 The transfer of information between estimating and other functions in a contracting organisation — or the case for going round in

circles. CIOB Construction Papers No. 19.

Harvey RC and Ashworth A 1993 *The construction industry of Great Britain,* Newnes.

Herzberg F 1966 *Work and the nature of man,* World Publishing Co.

Hillebrandt PM and Cannon J (eds) 1989 *The management of construction firms,* Macmillan.

Hillebrandt P, Lansley P and Cannon J 1995 *The construction company in and out of recession,* Macmillan.

Hofstede G 1984 *Culture's consequences* (abridged edition), Sage Publications.

Horner RM 1982 *Productivity, the key to control,* CIOB Technical Information Service, No. 6.

Hutton W 1995a How to prime the jobs generator. *The Guardian* 3 April: p. 13.

Hutton W 1995b *The state we're in,* Jonathan Cape.

Jackson MP 1991 *An introduction to industrial relations,* Routledge.

Katz RL 1971 Skills of an effective administrator. In Bursk EC and Blodgett TB (eds) *Developing executive leaders,* Harvard University Press.

Koontz H and Weihrich H 1990 *Essentials of management,* McGraw-Hill.

Lai Hon-Keung 1989 *Integrating total quality and buildability: A model for success in construction,* CIOB Technical Information Service, No. 109.

Langford D, Hancock MR, Fellows R and Gale W 1995 *Human resources management in construction,* Longman.

Latham Sir M 1994 *Constructing the team,* HMSO.

Lavender, SD 1990 *Economics for builders and surveyors,* Longman.

Lawrence PR and Lorsch JW 1967 *Organization and environment,* Harvard University Press.

Lewin K 1951 *Field theory in social science,* Harper.

Likert R 1961 *New patterns of management,* McGraw-Hill.

Lock D 1992 *Project management,* Gower.

Mace D 1990 Problems of programming in the building industry. *Chartered Builder* March/April: pp. 4–6.

Machlup F 1967 Theories of the firm — Marginalist, behavioural, managerial. *American Economic Review.* March.

Maslow A 1970 *Motivation and personality,* Harper & Row.

Mayo E 1945 *The social problems of an industrial civilisation,* Harvard.

McClelland DC 1961 *The achieving society,* Van Nostrand.

McGregor D 1960 *The human side of enterprise,* McGraw-Hill.

Megginson L, Mosley D and Pietri P 1992 *Management — Concepts and applications,* Harper Collins.

Mintzberg H 1973 *The nature of managerial work,* Harper & Row.

Monaghan TJ 1987 *Practical application of quality assurance to construction,* CIOB Technical Information Service, No. 80.

Moor N 1995 Room for manoeuvre. *Building* 23 June: pp. 22–3.

Morrison G 1995 Partners in quality. *Chartered Builder* April: pp. 9–11.

Murdoch J and Hughes W 1992 *Construction contracts law and management,* Spon.

Neale A and Haslam C 1994 *Economics in a business context*, Chapman & Hall.

Nicholls CJ and Langford DA 1987 *Motivation of site engineers*, CIOB Technical Information Service, No. 78.

Pateman J 1986 There's more to quality than quality assurance. *Building Technology and Management*, August/September: pp. 16–18.

Preston D 1995 Alive and kicking — Quality assured. *Chartered Builder* April: pp. 12–13.

Pugh DS and Hickson DJ 1993 *Great Writers on Organisations, the Omnibus Edition*, Dartmouth.

Ramus JW 1989 *Contract practice for quantity surveyors*, Heinemann Newnes.

Ridout G 1994a Survival of the fattest. *Building* 21 October: pp. 20–3.

Ridout G 1994b Plain selling. *Building Homes* 25 November: p. 23.

Schein EH 1980 *Organisational psychology*, Prentice Hall.

Scott S 1995 Record keeping for construction contractors. CIOB Construction Papers, No. 50.

Smit J 1990 Pivotal plant. *New Builder* 6 December: p. 29.

Stewart A 1994 For whom the banks role. *Building* 14 January: pp. 14–15.

Stewart A 1995 Homebuyers take new turn. *Building* 10 February: p. 18.

Stewart R 1983 *Choices for the manager*, McGraw-Hill.

Tannenbaum R and Schmidt W 1958 How to choose a leadership pattern, *Harvard Business Review*, March–April.

Thomas R 1994 Dividends inflated to repel takeovers. *The Guardian* 24 November: p. 20.

Tuckman BW 1965 Developmental sequence in small groups. *Psychological Bulletin*.

Walker A 1989 *Project Management in Construction*, BSP.

Willis CJ and Ashworth A 1987 *Practice and Procedure for the Quantity Surveyor*, Collins.

Wood S (ed) 1982 *The Degradation of Work? — Skill, deskilling and the labour process*, Hutchinson.

Woollacott M 1995 Weak firms and frail unions must link arms to survive. *The Guardian* 1 March.

Index